Research supported by Shanghai Sailing Program

Supported by the National Fund for Academic Publication in
Science and Technology

Mobile Microservices
Building Flexible Pervasive Applications

Nanxi Chen

人 民 邮 电 出 版 社
北 京

图书在版编目（ＣＩＰ）数据

移动微服务 ： 构建灵活的普适应用 ： 英文 ／ 陈南
希著. -- 北京 ： 人民邮电出版社，2022.4
书名原文: Mobile Microservices: Building
Flexible Pervasive Applications
ISBN 978-7-115-58402-1

Ⅰ. ①移… Ⅱ. ①陈… Ⅲ. ①互联网络—网络服务器
—英文 Ⅳ. ①TP368.5

中国版本图书馆CIP数据核字(2021)第275462号

Summary

This book focuses on the application level aspects of microservices, and addresses the combination of microservices in pervasive computing environments.

It introduces design concepts for pervasive applications with microservices, microservices deployment in edge/fog computing environments, microservice composition model, cooperative microservices provisioning to improve the overall service availability, and an implementation case study for microservices (GoCoMo).

The book also evaluates how the proposed solutions fulfill the identified challenges and research questions in pervasive computing environments, summarizes the research presented, and provides a list of interesting open issues that require further research.

It can serve as a beneficial reference for researchers and engineers in the area of edge computing, AIoT and 5G network applications. Graduated students in CS/EE areas will also find it useful.

◆ 著　　　　　Nanxi Chen
　　责任编辑　贺瑞君　赵　一
　　责任印制　李　东　焦志炜
◆ 人民邮电出版社出版发行　　北京市丰台区成寿寺路 11 号
　　邮编　100164　　电子邮件　315@ptpress.com.cn
　　网址　https://www.ptpress.com.cn
　　固安县铭成印刷有限公司印刷
◆ 开本：700×1000　1/16
　　印张：15.25　　　　　　　2022 年 4 月第 1 版
　　字数：299 千字　　　　　2022 年 4 月河北第 1 次印刷

定价：149.00 元

读者服务热线：**(010)81055552**　印装质量热线：**(010)81055316**
反盗版热线：**(010)81055315**
广告经营许可证：京东市监广登字 **20170147** 号

To my mother

Rui Chen

Preface

In the 5G era, edge computing enables new business models, strategies, and competitive differentiation, as with ecosystems of mobile microservices. This book targets pervasive application development on edge computing systems. It establishes concrete, technology-centric coverage with a focus on the concepts, architectures, well-defined building blocks, and prototypes for mobile microservice platforms and pervasive application development. Subsequent to the technology-centric coverage, the book proceeds to establish metrics that allow for the economic assessment of connected, smart mobile services. With the proposed model, costs can be minimized through statistical workload aggregation effects or backhaul data transport reduction, customer experience, and safety can be enhanced through reduced response time.

This book is a straightforward resource for getting started with microservices and explore the use of microservices in pervasive applications. It will take you through the challenges and requirements of developing an application in the computing environment of 5G. You will then move to the fundamental concepts of microservices and learn the way to make the right choices when designing microservices and applications based on microservices. As you progress, you'll be taken through the best practices while implementing and deploying your microservice applications. At the end of the book, you'll learn how to design, implement, and deploy a microservice-based pervasive application.

This text reveals the challenges of developing pervasive applications in the 5G era, focuses on how to build pervasive applications based on microservices, proposes a goal-oriented service composition model, and provide a complete implementation and configuration to build a service middleware and an AI-based microservice. Readers can learn about essential concepts, technologies, and tradeoffs in the edge computing architectural stack to help them building a pervasive application.

This work is sponsored by Shanghai Sailing Program (No.19YF1455900).

Contributors

Siobhán Clarke
Trinity College Dublin,
Dublin, Ireland

Yi Han
Wuhan University of
Technology, Wuhan, China

Jin Li
Technische Universiteit
Delft,
Delft, Holland

Yanbei Li
Shanghai Institute of Microsystem and
Information Technology,
Shanghai, China

Yang Yang
ShanghaiTech University,
Shanghai, China

Wu Yecheng
Sinolink Security,
Shanghai, China

John K. Zao
The Chinese University of Hong
Kong,
Hong Kong, China

Tao Zhang
National Institute of Standards and
Technology,
Gaithersburg, USA

Contents

Chapter 1

Introduction

The 5th generation mobile communication technology (5G) and edge computing have become important network infrastructure and are expected to be widely used in the coming decade. They not only tremendously increase the network capacity and capabilities, but also help to complete the convergence of computing and communications to make computing truly pervasive [1]. Pervasive computing environments enable access to diverse resources and services over networked computing systems. Such a system includes traditional computers as well as embedded devices, information appliances, and sensors [2, 3].

During the decade of 2010s, breakthroughs in achieving faster and more energy efficient wireless communication as well as smarter and thinner embedded devices have accelerated human users' shift away from traditional computers towards mobile and embedded devices for resource access and information sharing. The emergence of the Internet of Things (IoT) further enables the connections among massive diverse entities such as vertical systems, communication networks, smart devices/things, and applications. These entities enrich IoT with a plethora of data and various new applications. Thus, there is an increasing demand for developing pervasive applications that are feasible in 5G and edge computing environments.

Service-oriented computing is a major paradigm in pervasive computing [4, 5]. It packages heterogeneous resources as services or microservices that are discoverable, accessible, and reusable. It also provides unifying interfaces for microservices to ease users' access via communication networks. To meet a particular user requirement, a combination of multiple microservices may be required, and so a fully-functional service composition process will tackle complex user requests with a flexible composition of value-added services. This makes the microservice-oriented model an ideal way to create flexible applications and to tackle the mobility issue of the computing environment.

Besides the conventional services from the Internet industry, communications service providers have started to position themselves as partners in key verticals to secure their places as priority providers of pervasive applications [6, 7, 8]. Edge computing inspired many early attempts to combine communication services with traditional service computing platforms to support pervasive applications [9, 10, 11]. Edge computing pools the distributed computing resources to support applications, and service platforms will be anywhere along the cloud-to-things continuum, including in the cloud, at the edge of the network, or on the things [12]. This allows an end user to offload tasks from the cloud (or from the end devices) to edge devices that reside in the vicinity of the end users or the data sources, which will reduce the latency and bandwidth required for transporting data to the cloud.

This book establishes concrete, technology-centric coverage with a focus on the concepts, architectures, well-defined building blocks, and prototypes for mobile microservice platforms and pervasive application development. Subsequent to the technology-centric coverage, the book proceeds to establish metrics that allow for the economic assessment of connected, smart mobile services. In some cases, edge computing enables new business models, strategies, and competitive differentiation, as with ecosystems of mobile microservices. In other cases, costs can be minimized through statistical workload aggregation effects or backhaul data transport reduction, customer experience. Safety can be enhanced through reduced response time. Revenue and competitive advantages can also be enhanced through new edge-enabled service provisioning models.

Topics covered in this book include:
- The challenges and requirements of 5G and edge computing
- How to design flexible mobile microservice-based applications
- Microservice architecture and models
- Prototype development and examples for AI-based applications
- Performance and maintenance of microservice-based applications

This book focuses on the application level and addresses the combination of microservices instead of edge devices. Here's the content structure of the book.

Design Concepts for Pervasive Applications

Chapter 2 points out the service composition challenges with regard to the features of open and dynamic pervasive computing environments. It then investigates the state of the art and explores how it meets the challenges in the target computing environment. It applies an assessment metric to review the feasibility of these solutions and classifies them. Based on the analysis results, the chapter proposes a set of design concepts and shows how the design addresses the challenges.

Microservice Deployment in Edge/Fog Computing Environments

Edge computing allows microservices to be pushed down from the cloud to the edge devices, which enables low-latency services access between end users and edge devices. Service composition has an execution flow that requires multiple edge devices' participation. To reduce the latency of service access between edge devices, Chapter 3 focuses on the deployment of microservices in edge/fog computing environments to achieve the shortest service execution route and to deal with the adaptation issues that tackle dynamic environments.

Microservices Composition Model

Chapter 4 describes the design objectives of microservice composition and lays out the system model in detail. In particular, it begins with describing the requirements that should be satisfied and the trade-off based on the analysis in Chapter 2. It then describes the service composition model as a proposed solution to the problem of service composition in open and dynamic pervasive computing environments.

Cooperative Microservices Provisioning

Edge devices can be owned by heterogeneous third-party providers. There is a lack of a central management entity to offload service execution tasks. To optimize the distribution of microservices and increase the efficiency of service composition, Chapter 5 explores the selfishness of edge devices and proposes a game-based microservice provisioning model to improve the overall service availability.

Implementation

Chapter 6 describes the implementation of the above models and introduces a support middleware. It also presents two prototypes. A Java-based prototype realizes the proposed models as a middleware application. A C++ implementation integrates the proposed service composition model with the network simulator ns-3 for evaluation. It is realized as an extension module on the ns-3 platform. Chapter 7 demonstrates a detailed implementation for enabling intelligent services in a pervasive application at the network edge.

Evaluation

Chapter 8 evaluates how the proposed solution fulfills the identified challenges and research questions by comparing to baseline solutions from the state of the art. It begins with the introduction of evaluation metrics for a measurement of the proposed models and the baseline approaches in the target environment. Evaluation metrics include measurements of composition success rates under various mobility models, the composition model's scalability and performance, efficiency of service redeployment, and the availability of microservices. This chapter continues by introducing a prototype case study that demonstrates the proposed service composition model's feasibility on real mobile devices, and a simulation-based experiment. Simulation results illustrate both the strengths and limitations, under different network density and composite complexity conditions.

Discussions and Conclusions

Chapter 9 analyzes the achievements of mobile microservice research, summarizes this work and presents a list of interesting open issues that require further research.

Chapter 2

Design Concepts for Pervasive Applications

Pervasive computing environments have evolved from closed (special purpose) and static to open and dynamic (mobile) [3, 4, 13]. A large number of third-party mobile entities (e.g., wearable devices, smartphones, IoT devices, etc.) can be included in such environments. As modern wireless communication technology facilitates wireless data exchange for mobile users, various information captured by smart mobile devices like news, locations, air quality, reviews, routes/directions, and even parking/loading data, can be shared through wireless networks [14, 15, 16].

2.1　Motivating Scenario: A Smart Public Space

As a motivating scenario, Figure 2.1 shows a smart public space, which is a pervasive computing environment. A user issues a complex service request to a pervasive computing environment. Connected entities offer their hardware/software capabilities and local data as microservices. Anne is on a bicycle near a street, using her smartwatch to plan routes. She would like a 10-mile training cycle route that has less air pollution (See Figure 2.2). There is a local smart traffic system including various embedded devices owned by public transportation companies, weather service providers, taxi drivers, or pedestrians, etc. These devices can package their capabilities, like GPS, navigation, translation, real-time weather services, or city map to be accessible via network connections. Anne's smartwatch has been configured to incorporate surrounding communication networks to make use of available resources.

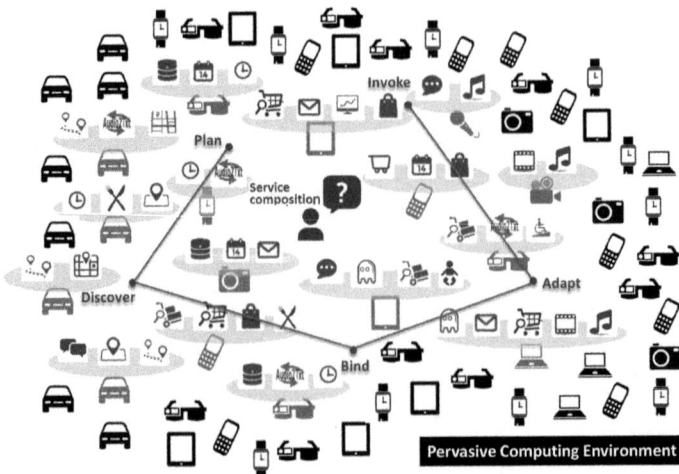

Figure 2.1　Motivating scenario: a smart public space system

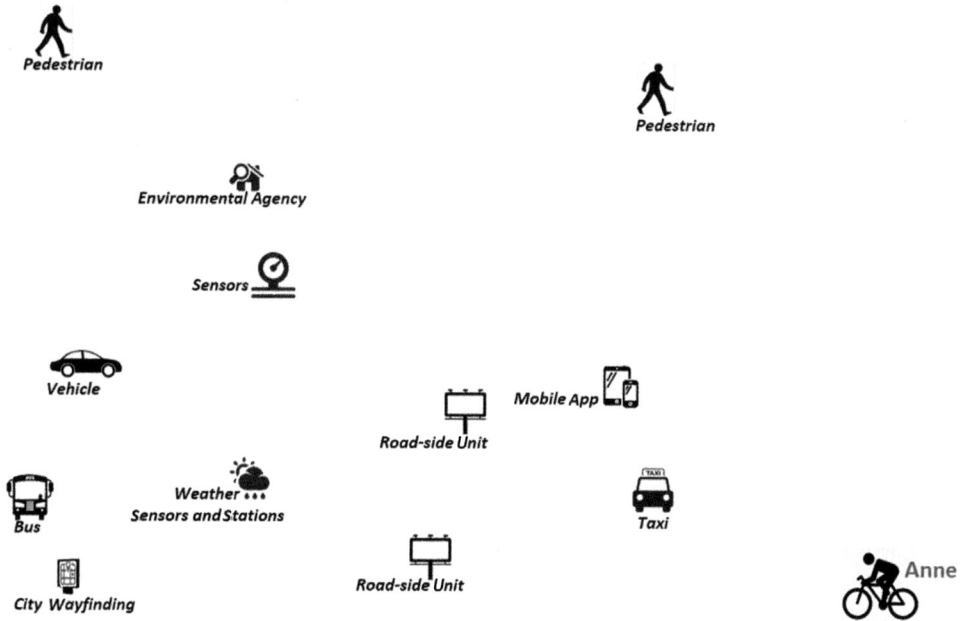

Figure 2.2   Anne's service composite

The smart traffic system is capable of allowing (possibly third-party) service providers to join the environment, discovering services based on the given requirements in a flexible way, producing a service workflow for data transition, and invoking service instances for execution. It has the potential to help Anne get the routing result she needs without browsing websites or utilizing the full capabilities of her hardware and software (See Figure 2.3). Thus, her smartwatch may be able to get a composite service (left picture in Figure 2.4) directly via a local network that forms by different devices in the pervasive computing environment, such as a road-side unit, a nearby car's satellite navigator, an environmental agency's sensors, etc. The composite service allows Anne to input only her current location and speed, and returns a 10-mile route that is less polluted. The service also returns an audio route stream that compacts with her smartwatch, guiding her through the training route.

This smart public space has the potential to help Anne avoid browsing websites, increasing the likelihood of matching all her hardware and software capabilities to get the routing result she needs. However, Anne's requirements are complex and domain-crossing such that an exactly defined service composition task may not be flexible enough to facilitate successful matching between her requirements and the available services in the environment. To enable service provisioning in such a smart public

Figure 2.3　A composite service to satisfy a complex task

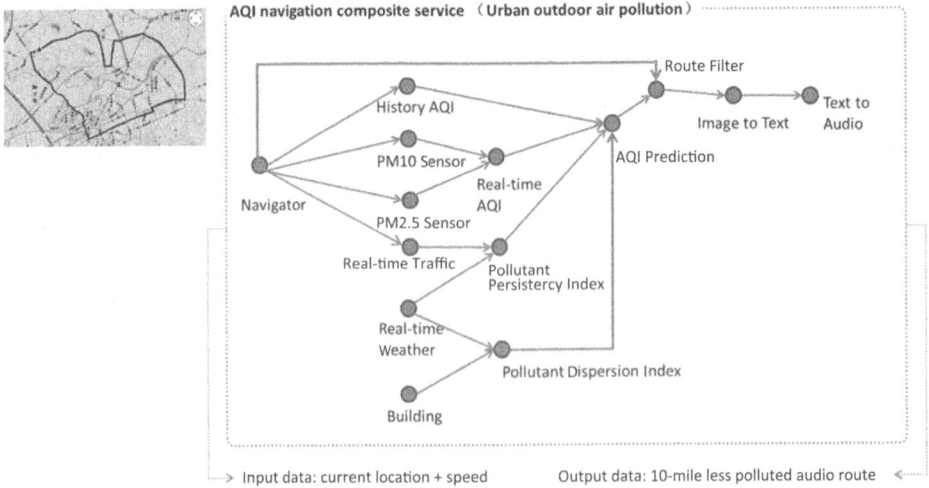

Figure 2.4　Input and output data of the composite service

space, a system should be capable of allowing (possibly third-party) service providers to join the environment, discovering services based on the given requirements in a flexible way, producing a service workflow for data transition, and invoking service instances for execution. Moreover, participant devices that host microservices are likely owned by heterogeneous third parties, such as communication service providers, pedestrians, taxi drivers, or weather agencies, and these device owners' activities are not under obligation to the system. The participant devices, therefore, may leave the system or drop the service before execution. This book assumes that services deployed on participant devices are stateless. For successful service provisioning, the system should support timely service invocation, and be aware of and engage newly entered devices that have the potential to offer the given functionality throughout the service composition process. Figure 2.5 shows the typical process of microservice composition.

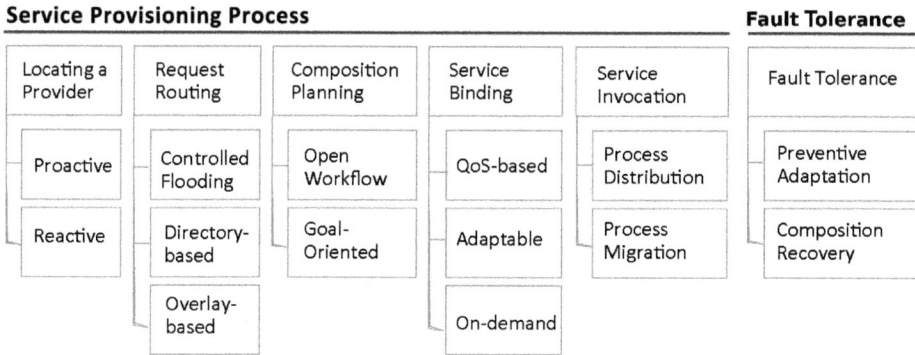

Service Provisioning Process					Fault Tolerance
Locating a Provider	Request Routing	Composition Planning	Service Binding	Service Invocation	Fault Tolerance
Proactive	Controlled Flooding	Open Workflow	QoS-based	Process Distribution	Preventive Adaptation
Reactive	Directory-based	Goal-Oriented	Adaptable	Process Migration	Composition Recovery
	Overlay-based		On-demand		

Figure 2.5   Process of microservice composition

2.1.1   Challenges

Given the environment's openness and dynamism, service composition processes face significant challenges. The primary challenges are listed as below.

1. Inadequate Conceptual Composites

A conceptual composite is an abstract model that states a specific functionality as a series of service requirements, each of which indicates that a concrete microservice will be used. General service composition relies on previously generated conceptual composites and assigns service providers at runtime to complete them. However, in an open environment, microservices of interest are independently deployed and maintained by different edge devices so that a conceptual composite for a particular functionality is not always possible. Composition users may provide the conceptual composite as a part of the service request, but such a user-defined composite is likely to be at variance with the operating environment as the environment is dynamic. In addition, a composition user's awareness of available microservices is limited by its communication range, so the composition user may have insufficient knowledge about usable microservices and cannot produce a conceptual composite for service composition. A predefined conceptual composite also removes the possibility of using microservices that may contribute to the user's request, but are not outlined in it.

2. Limited System Knowledge

Service discovery searches for and selects services that match the required functionality. A service provisioning system may need to dynamically reason about a functionality and find an appropriate combination of microservices that supports it when an individual one is unavailable. Having a global system view of the computing environment is beneficial for such a reasoning-based service discovery, since global service knowledge will facilitate reasoning processes and increase composition success. However, obtaining service knowledge from mobile devices leads to traffic overhead because it relies on multi-hop wireless data transmission, and when edge devices have a limited communication range, there are likely to be a larger number of transmission hops. In addition, maintaining a global system view can be expensive especially when the service and network topology are frequently changing.

3. Unpredictable Service Availability

Any microservice provider can offer and drop services, as well as join and leave the pervasive network at arbitrary times during execution, making services' availability unpredictable at runtime. Service execution can fail due to a previously available microservice's absence. New service providers can enter the network and offer new microservices, which brings significant potential for improving overall service quality by re-composing better microservices from the environment, including those that may have appeared even during service execution.

4. Unreliable Wireless Communications

Many edge devices rely on wireless communication channels to exchange service information, bind services, and transfer data during service execution [4, 17]. Wireless communication is likely to be unreliable, and the data packet transmission via the radio channel is sensitive to interference and obstruction, with packet loss coming as a result [18]. Efficient interactions between composite participants are required to reduce the dependency on such a communication channel.

5. High Level of Dynamics in Execution Paths

Service execution requires communications through established wireless links between successive service providers. However, service providers' and service consumers' mobility leads to network topology changes, which alters such links. Changes to the links make any cached execution path error-prone, which may further result in communication failures and execution path loss during service execution. A composite service must be adaptable to increase the chance that results can be delivered even when a communication channel to a provider drops, or a cached execution fails. Recovering the composition from a failed path in a time and communication efficient way is also required to achieve a successful service composition.

2.1.2   Possible Solutions

Service composition's success in open and dynamic environments is subject to failures that can occur in any phase of the composition process. Support for services provision using composite services in such environments should include both composition process management and failure recovery. Composition process management investigates different ways to organize a composition process to increase composition success. Fault tolerance explores methods to predict potential failures and adapt the composite service for them, or to recover a service composition process from an emerged failure.

1. Composition Process Management

A service composition process is initiated by the service provisioning system when a composition request is issued. The process refers to the tasks of reasoning about a composition plan, discovering available services that match the plan, binding the discovered services and invoking service instances. Challenges when designing a service composition process include the questions of when and in what order these tasks are performed, which entity in the network executes the tasks, where the service provisioning system retrieves necessary knowledge to support such execution, and how these tasks are managed. The state of the art for service composition in pervasive environments includes runtime composition reasoning [19], decentralized service composition [20], dynamic service binding [21, 22], weaving the service invocation procedure to the service discovery procedure [23], etc., which cope with flexibility and context dynamism. There are existing solutions adopt or expand these techniques. Main techniques that address the challenges of the target system are planning-based service composition and decentralized composition management.

Planning-based service composition differs from traditional matching based service composition approaches that rely on a given conceptual composite that is made up of a set of ordered abstract services, and only finds service instances that exactly match the abstract services. Planning-based service discovery is not restricted to exactly matching a composition task, making it more flexible with the potential to a broader scope of services than matching-based service composition. Planning based service composition solutions extend conceptual composites or use AI planning algorithms to increase flexibility. Conceptual composite extensions use a one-to-more matchmaking scheme for service composition, which means one abstract service may match a combination of services [20]. However, this model still requires a conceptual composite that is planned based on offline service information, which may be out of date. On the other hand, AI planning algorithms are goal-oriented, automatically creating abstract composites (composition tasks) during service discovery, selecting services that match each composition task, and finalizing the composites by invoking selected service instances at runtime. Some of them can support various service workflows other than sequential workflows but require planning infrastructures [24, 25, 26] or a global knowledge base [25, 27].

Decentralized composition management [21] distributes a conceptual composite to different participants, each of which carries out a part of the abstract services. In general, overall service invocation starts only after every abstract service in the composite has been bound to a provider [28]. This protects a composition from unnecessary invocation and execution when some of the required providers are non-existent. However, even when each required provider has been located and bound, the composition can still fail because of absent service providers at runtime.

Minimizing the impact of service availability has been explored by employing on-demand service binding [23]. On-demand service binding assigns a provider to an abstract service just before it has to execute. Existing approaches have explored on-demand service binding in two directions: one discovers a set of providers for every required service as allocation candidates and maintains these candidates until one of them is invoked for execution, and the other discovers a provider at runtime and invokes its service immediately after the provider is found. However, using allocation candidates [21] requires additional monitoring effort to keep the list of candidates up to date, and limits the chance of engaging better providers that may appear in the environment after all the allocation candidates for service are determined. On-demand service binding [23] can obtain real-time service provider information by runtime service discovery. It reduces composition latency by coupling the composition process with the invocation of services. However, the on-demand service binding approach relies on exactly matching functionalities and predefined conceptual composites for service discovery, which reduces flexibility. Moreover, it assumes broadcast-based service discovery, and provides no means to prevent a service composition request from flooding the network. Furthermore, it migrates composition processes along the direction of data flow, which may mean that the system invokes a final provider a long distance from, or a bad transmission channel to, the user. This increases the cost of routing the composition result back to the user. In summary, there is a lack of service composition process to resolve clients' diverse composition requests, by using only participant providers' local service information. Its service discovery method should be infrastructure-free, aware of new service provider that newly emerges, and can control discovery message flooding to reduce congestion in the network.

2. Fault Tolerance

Research on fault tolerance in service composition focuses on failure prevention and composition recovery. Failure prevention approaches tend to anticipate composition failures and update the composite service, which requires continuous observation of changes in the operating environment throughout the overall composition process and the adaptation of the composite to observed changes [29]. Observations on the operating environments are normally achieved by sending probe messages to composite participants to assess the allocated services' availability and the validity of links between service providers [21].

As the operating environment is highly dynamic, an early failure prediction tends to be inaccurate. Such a prediction is likely to lead to unnecessary adaptation. On the other hand, a just-recently-detected impending failure leaves very limited time for a composite service to update itself. Failure prevention may fail if a composite has not been completed yet.

Observations: Traditional solutions for service composition in pervasive computing have explored composition process management and fault tolerance to address openness and dynamism. As mentioned above, composition process management solutions include efficient service discovery [30], flexible planning [20, 24, 31], dynamic binding [23], decentralized composition management. Most planning-based service composition approaches rely on discovery infrastructures and a-prior system knowledge. Decentralized composition management solutions are restricted to exact matching functionality or planning for a sequential service composite. Fault tolerance either requires composition re-planning that is time-consuming or only replaces failed providers with backup ones without considering the reliability of the new execution path. Taking full advantage of the mobile and wireless technology advances and opportunities in open, dynamic pervasive computing environments will therefore require the development of new and more appropriate service composition models that provide Flexible decentralized composition of services on mobile devices, as well as Dynamic adaptation of the combination of services appropriate to the service provider and the environment's changing situation.

Flexible decentralized composition of services on mobile devices means the target network that is established by ad hoc connected mobile devices is infrastructure-less, requiring that the process performs in a decentralized manner. Models are also required to streamline to achieve efficient composition.

Dynamic adaptation of the combination of services appropriate to the service providers' and the environment's changing situation means, where there are constraints such as timeliness, differing data flows, combinations of services will be appropriate under different circumstances. Models are required to quantify services' and their combinations' characteristics (e.g., service availability, execution path reliability) and to adapt as appropriate to environmental factors.

2.2 Locating A Provider

Service providers specify their local capabilities as microservices and make the service specification available to service provisioning systems. Service discovery includes active search and passive search [17] ①. Reactive discovery (a.k.a., passive search) means that a service provisioning system searches microservices on an entity (or a set of entities) that previously receives providers' announcement and caches their service specifications, and proactive discovery (a.k.a., active search) model directly enquires service providers for their service specification.

Specifically, **reactive discovery** requires service providers to register service specifications to the system by periodic service advertisement. Networked entities in the system cache such advertisements and manage them in a particular structure for quick queries. Decentralized caches are attractive in pervasive computing environments, as they better suit mobile environments compared to their centralized counterpart [32, 33, 34] that assumes the presence of a resource-rich, stable repository to cache all the providers. **Proactive discovery** allows providers to respond to a composition request by replying with their dynamic service specification to the system.

This book considers the process for locating providers from two perspectives: how to obtain dynamic service information for composition planning, and how to get current service information in a timely manner when service re-composing is needed to recover a composite service from failures.

---

① "Active search registers microservices in local cache, and search takes place in all microservices hosting devices by flooding the search message. Passive search registers their microservices in all nodes, and search takes place in only local cache [17]".

2.2.1　Reactive Discovery

In Distributed Service Discovery Model (DSDM) [28], a set of devices are selected as service directories that store registered providers' information, including provider addresses, the description of service capabilities, and QoS values. Each directory keeps a copy of this information for every registered service provider. As a consequence, any of the registered service providers' information can be directly fetched by a service requester from the nearest directory. However, such duplicated directories make the network's maintenance cost exponentially increases with a growth in number and density of providers as all the directories have to be refreshed, which results in limited scalability.

Hierarchical service composition [35] classifies service providers into four different levels according to their resources and computing capability. Co-located service providers are managed as a hierarchical overlay network. Resource-abundant providers are ranked as high-level and act as parent nodes in this hierarchical network. Parent nodes maintain their offspring nodes' I/O dependency as a graph, and each of the offspring nodes has limited resources and keeps a subset of the graph. Parent nodes have more system knowledge than their offspring. In this hierarchical model, a system looks for services in the available graph subsets in the service requester's close vicinity and traverses the overlay network upwards if the nearby graph subsets have inadequate information to satisfy the service requirements. This model allows networked service providers to maintain a part of the system knowledge depending on their capability. Keeping system knowledge on each service provider up-to-date relies on receiving heartbeat messages from neighbouring service providers to identify their presence. However, updating this hierarchical overlay network may not only need to modify a local service dependency graph but also to upload the graph, to parent nodes, which increases network traffic.

The work of Sadiq et al. [36] facilitates service composition with load-awareness and mobile awareness. In this model, service providers publish their services to networks in their vicinity. Each participant service provider maintains a service I/O dependency graph that includes information about neighbouring service providers, the current load of these providers, and the temporal/physical distance between services. Based on the dependency graph, services with the shortest temporal distance will get selected to form a composite service. This model supports distributed global knowledge, by which a service composition system uses multiple services I/O dependency graphs stored in different service providers to resolve a composite request, but in this model there's a propagation delay that may outdate service information.

The work of Daneshfar et al. [37] tackles the uncertainty of resource availability in mobile networks. It models the execution cost of a given set of services and introduces an integer optimization formulation to minimize the total cost of service provisioning. To do so, it takes trade-off of the service cost and the QoS requirements of users, and introduces a factor of "probability of availability" that captures the mobility and its effect on the QoS to achieve the model's goal.

Cao et al. [38] proposed a renewable-adaptive approach to assign services according to the available energy of their service hosts. They introduced a QoS function for an application, which is expressed as the total amount of successfully transmitted output data. The system selects the service provider with a high QoS through a game theory-based algorithm with the goal of maximizing the QoS of not only each application but also the entire system.

2.2.2 Proactive Discovery

The work of Prinz[21] relies on potential service providers to respond to service requests with their execution properties and to mark themselves as candidate providers. The request source selects the best performing service provider as the primary provider and expects the other candidates to monitor the execution of the primary one. This model has no need for service rediscovery to recompose services for failure recovery. Once the primary provider fails, the second-best performing candidate will take on the execution, and be monitored by the rest of the candidates. This solution uses a Distributed Hash Table (DHT) to locate potential service providers, so it can prevent expensive directory maintenance. In a DHT-based service discovery system, when a service provider joins the system, it gets assigned a key (hashing value) that corresponds to its service functionality. The system hashes the key and appoints a node in the system as an index to cache the service provider's information. Functionally equivalent providers tend to be given the same hashing value, so their information is normally stored in the same index node. However, this can make a DHT-based service discovery system brittle in dynamic environments as the system will just lose the possibility of providing functionality when the corresponding index node departs.

Recent research explores extensions to DHT-based service discovery systems to retain their lightweight approach for maintenance while preventing functionality loss. The work of Kang et al. [39] introduces an optimal search that integrates a vicinity service index overlay into a DHT overlay network. It is not only the index node but also its neighbouring nodes that keep service provider information for a hash value. The work of Pirrò et al. [30] combines DHTs with a semantic overlay network, which takes microservices that have semantically similar functionality to the requested one into consideration when searching for a service provider. It compromises the problem of functionality loss by expanding the service discovery scope, but it comes at the cost of maintaining the semantic overlay network. Chord4S [40] distributes the descriptions of functionally equivalent microservices to varying directories to maintain continuously discoverable functionality in open Peer-to-Peer (P2P) networks.

P2P-SDSD [41, 42] enables dynamic service collaboration in P2P networks. It applies a probe-based service query, and a service request is submitted to service providers through previously established semantic links. Semantic links categorize service providers by connecting microservices that share similar functionality and get updated when a service provider joins/leaves the category or enables/drops microservices. Service providers in this model cache the categories, based on which a group of providers for a specific functionality can be quickly located.

Service Specific Overlay Network (SSON) [20] composes networked entities in a decentralized self-organizing manner and addresses seamless media delivery in pervasive environments. SSON requires no previously established discovery infrastructure. However, service providers are assumed to be aware of their geographic locations, and a service request is forwarded to providers' physical neighbours with a specific geographic angle②. In dynamic environments where devices' geographic locations can change quickly, frequently updating of geographic location wastes local computation resources.

A cooperative discovery model [24] provides fully distributed support for unstructured P2P networks. With this model, a service provider receives a composition request and updates it by removing the parts of functionality that can be supported by the provider. Then, the service provider adds itself into a solution table and forwards it along with the updated request to other providers. This process continues until all the required functionality cleared and a set of service providers recorded in the solution table. This model can be built over an existing overlay network (e.g., semantic overlay network or DHT) to improve search efficiency. The work of Furno [24] additionally proposes an

---

② "...a geographic angle [0, 180] determines the search scope between the media client and the media server [20]...".

overlay to further increase the search efficiency, which will be discussed in Section 2.3.3.

Opportunistic service composition [23], similar to SSON, assumes no infrastructure for service discovery, it broadcasts composition requests and locates a service provider on-demand. It introduces a cross-layer service discovery that extends traditional broadcast protocols with one-hop acknowledgement messaging. With this cross-layer service discovery model, a service provider is aware of its neighbouring service topology. This topology information enables multicast-based service binding and unbinding which reduces network traffic.

CloudAware [43] targets the resource constraints in mobile computing environments and presents a context-adaptive middleware to adapt to the context change. It implements a lightweight data mining process to predict the future availability of WiFi, a bandwidth range, and realvalued attributes like the execution time of an invoked service. Such forecasting of a future value of a certain context attribute is enabled based on historical data collected by mobile devices. Adaptive Fog Configuration (AFC) [44] also aims to solve the resource constraints of service hosting nodes and maximize the utility of resources, it jointly optimizes service hosting and task admission decisions by predicting service demands. It is realized based on Lyapunov optimization and parallels Gibbs sampling, requiring only currently available system information.

2.2.3 Planning-based Composition Announcement

In general, a service discovery system can optimize runtime service search by using the previously cached service information (reactive discovery). This reduces communication overhead and discovery delay when locating service providers, but the system has to keep such service information up to date. Some approaches rely on service providers' advertisements to update service information. If such advertisements are sporadic, the providers' information may be out of date, and if they are too frequent, the energy cost of providers increases and in turn leads to a negative impact on providers' availability. Therefore, reactive service discovery is less efficient in terms of dynamic maintenance. However, service advertisement facilitates service composition in dynamic environments. A system may need to recompose better microservices from the environment, including those that may have appeared even during service execution, aiming to recover from composition failures or improve overall service quality in real-time microservices provisioning. Thus, the system has to be context-aware. Service advertisement at runtime has the potential to cut down monitoring effort for a composition handler when detecting new, emerged microservices, and does not require to consider the maintenance problem.

Proactive service discovery gets the dynamic provider information directly from service providers but may lead to high latency searching. Although there are overlay networks to improve the search speed, high latency searching still exists, especially when a composition requirement is complex. As service execution usually starts when every requested service in a composition has bound to a provider, a delay when searching for a service increases the potential that recently available providers may no longer be available. A complex composition requirement implies more required services. If a discovered provider has left the network before all the requested microservices are found, using this provider's information to finish the remainder of composition processes causes faults. Opportunistic service composition [23] assumes a sub-service in a composite service can execute as long as its required data is provided regardless of whether all the microservices are bound to their providers. It uses a service provider once it is located, which reduces the searching delay, but this approach is inflexible in other composition aspects like service composite adaptation that will be discussed in the following sections.

Service discovery approaches that rely on proactive search are efficient at obtaining current service provider information. The approaches usually require service providers to announce the services they offer. Proactive service discovery distinguishes two methods in terms of how they distribute composition request issuers: semi-decentralized discovery employs a group of composition brokers who issue requests and accumulate information about service providers that have the potential to support a composition process, fully-decentralized discovery considers each service provider as a request issuer that discovers providers for its subsequent microservices.

Figure 2.6 shows the service composition process of a sequential composite. Scale marks on the vertical lines show the number of interactions for service discovery and execution (the fewer communication, the better). The dash half-braces show the distance between successive service Providers (the near locality, the better). Figure 2.6 (a) and (b) illustrate the composition processes of semi-centralized service discovery [21, 45] and decentralized interleaved service composition [23] for a sequential composite, respectively. Note that Figure 2.6 depicts the interactions only to discover and invoke the primary service providers, those to find other candidate service providers and to bind service providers are omitted for simplification.

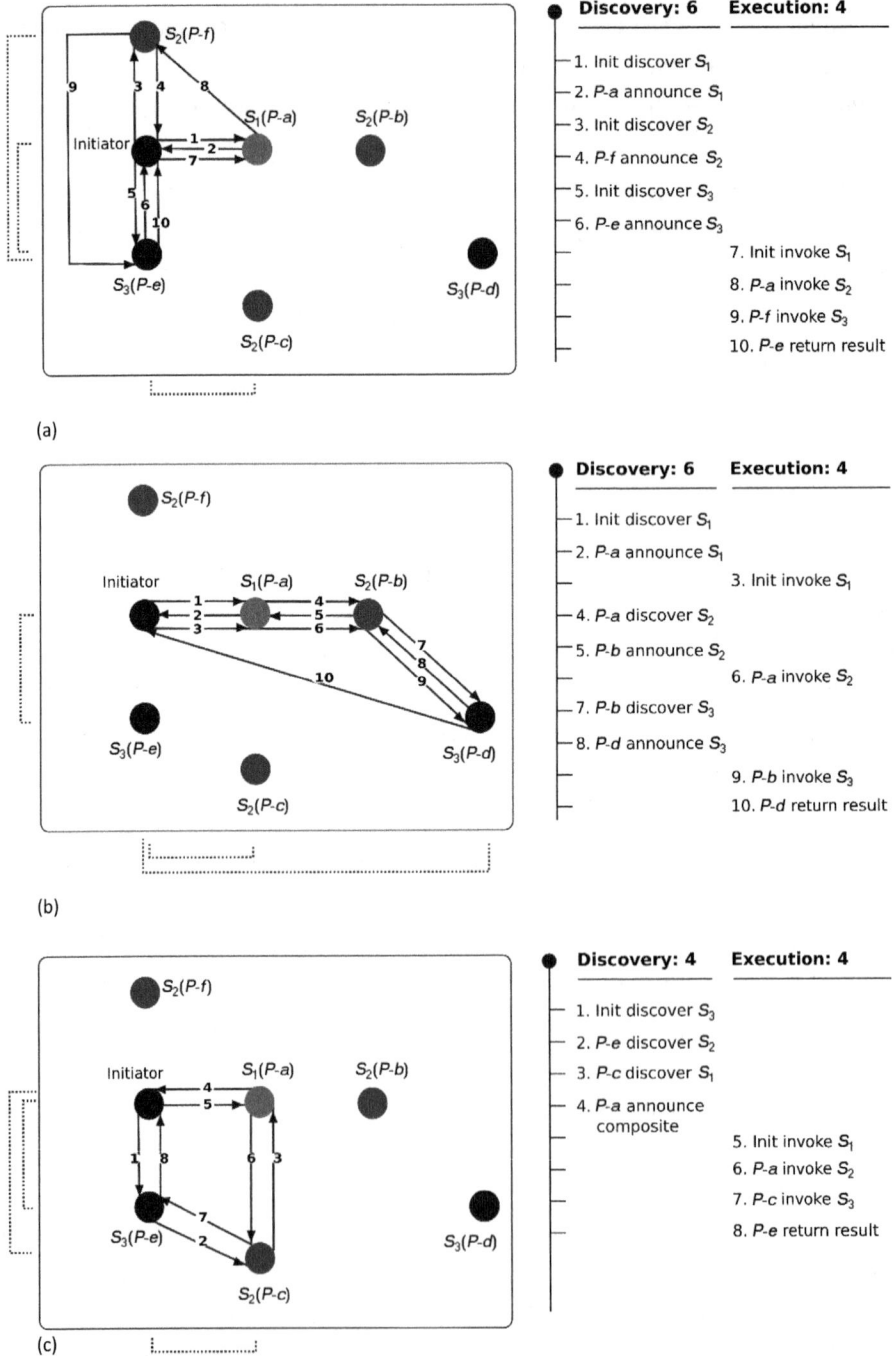

Discovery: 6   Execution: 4

1. Init discover S_1
2. P-a announce S_1
3. Init discover S_2
4. P-f announce S_2
5. Init discover S_3
6. P-e announce S_3
7. Init invoke S_1
8. P-a invoke S_2
9. P-f invoke S_3
10. P-e return result

(a)

Discovery: 6   Execution: 4

1. Init discover S_1
2. P-a announce S_1
3. Init invoke S_1
4. P-a discover S_2
5. P-b announce S_2
6. P-a invoke S_2
7. P-b discover S_3
8. P-d announce S_3
9. P-b invoke S_3
10. P-d return result

(b)

Discovery: 4   Execution: 4

1. Init discover S_3
2. P-e discover S_2
3. P-c discover S_1
4. P-a announce composite
5. Init invoke S_1
6. P-a invoke S_2
7. P-c invoke S_3
8. P-e return result

(c)

Figure 2.6   Service composition for a sequential composite $S_1 \rightarrow S_2 \rightarrow S_3$ (a) Semi-centralized, (b) Decentralized interleaved, (c)Decentralized backward planning-based announcement

The proposed mechanism shortens the routes in a service execution path and reduces the communication by minimizing the geographic distance in one-hop routing and the number of hops in multi-hop transmissions, named decentralized backward planning-based announcement, as Figure 2.6 (c) shows. The mechanism introduces the concept of **planning-based composition announcement**:

The proposed protocol removes the announcement for each individual service and uses a dynamic composition planning process during service discovery to allow a composition initiator to directly gain information about service composites from service providers. If a service provider receives a composition request and discovers that it partially fits the request's requirement, instead of announcing its own microservice to the requester, it forwards the parts of the request that cannot be satisfied by its local knowledge to other service providers. Service providers do not contact the requester unless they can completely fulfill their received composition request. The composition initiator is only informed of completed service composites from service providers and selects the most reliable (shortest routes and execution time) microservice composite for execution.

As illustrated in Figure 2.6 (c), the proposed protocol has less interaction between primary service providers and finds the shortest execution path. This protocol is based on a dynamic composition planning algorithm to resolve composition requests.

There are various different planning mechanisms for dynamic composition planning, such as graph-based planning[31, 46, 47], decentralized composition planning[20, 24, 48], etc. Given the lack of centralized infrastructure on target network, decentralized composition planning has the potential to support the planning-based composition announcement. Decentralized composition planning includes forwarding planning [20, 48], backward planning [48], and bi-directional planning [24]. However, existing approaches search for services depending on pre-established semantic service overlay networks [24], assuming service providers have local geographic information [20], or simply flood a request to the network [48]. These methods are expansive in mobile networks. In addition, they only reason about sequential service flows. The proposed service discovery model uses controlled flooding and backward planning. It extends these two methods to reduce network traffic and to support complex compositions, for instance, a service composite including parallel or hybrid microservice flows.

2.3   Request Routing

A composition request needs to be routed to the networked entities that can resolve the request and handle service composition. Existing solutions are classified as controlled flooding, directory-based routing, and overlay-based routing.

Controlled flooding prevents broadcast-based requests from flooding the network using imposed limitations, such as the maximum propagation limit for routing. Directory-based routing manages service providers' information in a set of directories. Depending on the size of the network and the density of available microservices, a central directory or multiple directories can be applied. This section discusses decentralized directories and the question of how a request is routed in a network of directories. A centralized directory [49] has obvious limitations in wireless environments as the scale of the network is restricted to the directory carrier devices' wireless communication range. Overlay-based routing allows service providers to link to each other through a particular connection (e.g., semantic similarity, dependency, etc.) between them. Thereafter, a service request can be transmitted via these links. This greatly reduces network traffic compared to limited flooding. Some overlay-based routing approaches also employ directories to increase search success or search efficiency. They are classified as hybrid routing. Request routing approaches are analyzed with regard to the question of how to find service providers in a timely fashion, and the cost of the discovery process.

2.3.1 Controlled Flooding

Group-based Distribution Service Discovery Protocol (GSD) [50, 51] is a model can be used in pervasive computing environments. Service providers assign a service group to each of the microservices they provide according to service functionality, using a domain-specific tree-like ontology. They advertise their microservices and corresponding service groups to their vicinity networks. The service information is cached only by their direct neighbouring nodes, and the service groups are stored by all the nodes in the vicinity. Such cached service groups can improve service search efficiency, as a node can quickly know if the required functionality can be supported by its local network according to the cached service groups. If a service request matches a cached service or a cached service group, the request can be directly routed to the service provider. If there is no matched service group, the request gets forwarded to the rest of the network. Moreover, to prevent a request flooding the network, a hop-count parameter is used in service requests, which is initialized by a request source and indicates the maximum number of request forwarding permitted. A web service discovery model [52] for dynamic environments employs a Time to Live (TTL) parameter to limit request flooding by messages' transmission time. This is more flexible than GSD, but it only considers the transmission cost for discovering a single service.

Opportunistic service composition [23] is based on broadcasting for service discovery, but it is not like GSD that uses service advertisement of previously cached service information and build directory networks. Instead, opportunistic composition employs acknowledge messages to collect local network's topology information while routing request messages. A composition requester can know its communication routes to all the candidate service providers through receiving an acknowledge message from them. The composition requester uses the routes information to release③ the candidate service providers that will not be executed.

Reduced Variable Neighbourhood Search (RVNS) data processing framework [53] applies a limited flooding mechanism that relays messages in a geographical rectangle residing between the service consumer and the service provider or in a predefined logical range. However, there is still a risk of flooding the network with request messages and additional communication overhead introduced by acknowledge messages involved.

2.3.2   Directory-based

The work of Kozat et al. [54] distributes a service directory, which achieves scalable, efficient service discovery. This work introduces a virtual backbone layer that organizes service directories as a P2P overlay network and builds it above the general service overlays. If a provider registers its microservice to the network, its registration message will be forwarded to every directory in a virtual backbone layer. In other words, a service specification is duplicated and cached by different directories. Given that the computing environment is dynamic, the service information kept in the virtual backbone layer must be renewed when new providers are registered. The virtual backbone layer uses asynchronous updating, which means directories in the layer are asynchronously renewed by spreading registration messages. A service discovery process depending on such a virtual backbone layer requires a service request to flood the backbone layer until a matching service provider is found. Updating of directories has

---

③ Opportunistic service composition requires service providers to lock their resources for composition as long as they have received the request of the composition and have committed themselves as candidate providers.

to keep up with network changes. However, if the network frequently changes, keeping the whole directory-based virtual backbone up to date can be extremely expensive.

Likewise, Hexell[55] applies networked directories to maintain a virtual backbone layer, but without requiring all the directories to keep a copy of any individual service specification. Hexell divides an environment into hexagon cells, the size of which is determined by wireless devices' communication ranges. The service provider that is closest to the center of a cell is selected as a directory. A selected directory caches information about the service providers that are currently in its responsibility cell. Similar to GSD[50, 51], Hexell groups microservices using a hierarchy ontology, and organizes microservices according to their groups in a directory. A directory updates and informs the rest of the directories about available groups in its responsibility cell. As a result, a directory located in another cell knows what functionality can be extracted from service providers in each cell through the directory-based virtual backbone layer, and can support global service queries for service requesters in its own responsibility cell. This group-based mechanism allows Hexell to update the directory layer potentially less frequently than the network changes, because there is no need for a directory to update the directory layer if a new microservice is functionally similar to a microservice that has already registered to the directory. Hexell's hexagon cells also provide a service discovery system with a clear view of the communication distance between service providers through the number of cells a communication crossed, by which a communication-efficient composite of microservices may be selected. However, a directory node is mobile, and when it is roaming, it is likely to leave the center of its responsibility cell or even move to another cell. Hexell provides no clarification on how this will affect the service discovery process and how to manage it.

Volunteer-based Service Discovery (VSD)[56] encourages resource-rich devices to volunteer for directory duty. Directories are isolated in VSD. They are known mutually only when they have co-registered microservices. A link is built between directories when they know each other's presence, and service requests are forwarded to directories based on such links. VSD also has the issue of directory loss due to mobility. Similar to VSD. The work of Kalasapur et al.[35] considers resource-rich devices as directories. As described in Section 2.2.1, directories are maintained in a hierarchical way, in which service discovery requests bottom-up travel to find a service composite.

With the widespread use of RESTful services and the development of data mining and Natural Language Processing (NLP) technology, directory-based service discovery enters a new stage that targets the RESTful services and applies data mining and NLP algorithms to improve the efficiency and accuracy of service discovery. Zhang et al. [57] employed a topic model to assign services to different topic groups based on the service goals extracted from semantic service description, which turns the service discovery problem into a topic search problem. They further improve this solution and proposed a service discovery framework [58] that can better model the user's functional goal. Similarly, Tian et al. [59] targeted the semantic feature representation of services and introduced a neural topic model. They use word embedding to map the text of service description to the same feature space and cluster the services based on their semantic feature.

However, resource-rich nodes are not always possible in mobile environments, which makes VSD [56] and the hierarchical directory solution [35] feasible only in some particular scenarios. The clustering-based service discovery methods [58, 59] require centralized service registration, which makes them mainly used in the Web Service domain.

2.3.3 Overlay-based

Using DHT-based overlay networks (See Section 2.2.2) for service discovery has been investigated in pervasive computing. In a DHT-based overlay network, a hash value is calculated for the requested functionality. According to this hash value, a destination to route the request is determined. The destination is a node that previously cached information about the service providers that support the requested functionality. The destination node receives the request and forwards it to these service providers using multi-casting. However, as analyzed in Section 2.2.2, DHT-based overlay networks store information about functionally equivalent providers in the same node, so functionality loss may occur if a directory departs or crashes. In addition, even though DHT-based overlay networks prevent service requests from flooding a service providers' network, it can end up with a long routing path, which in turn delays the request to be routed to potential service providers. The work of Kang et al. [39] integrates service caches to a DHT overlay network, which removes a part of a redundant request route on the physical network layer, and increases time-efficiency for DHT-based service discovery. But this comes at a cost of maintaining service caches.

Semantic Overlay Network (SON) [60] has been proposed for content-based search optimizing in P2P networks. The core notion of a SON is to bunch similar peers, meaning query processes can discover bunches instead of individual peers for faster locating. Semantic Driven Service Discovery in a P2P scenario (P2P-SDSD) [41, 42] introduces inter-peer semantic links into SON, which supports fine-grained (i.e., not a composite) functionality to be provided and more possible compositions of microservices. P2P-SDSD supports two policies for request routing: minimal and exhaustive. With the minimal policy, a service request is forwarded through semantic links in a SON until a matching service provider is located for the request. The exhaustive policy is designed for finding a service provider with the optimal non-functional features (e.g., reliability) among a set of candidate providers. To achieve this, a system stops request routing only when all the candidate providers are discovered. In short, the minimal policy is efficient, but cannot find an optimal result, while the exhaustive policy may support the composition of services with good QoS, but it needs to route a request to all the potential providers, which is inefficient.

A cooperative discovery model [24, 61] uses historical composition processes to construct a tree-like overlay network to improve the efficiency of service discovery. The overlay network is formulated by networked super-peers. For each historical composition process, one super-peer is elected among the participant service providers to cache these participants' information and the composite service's functionality. Thereafter, this super-peer responds quickly to service requests with similar functionality to the composite service it maintains. If a request can be solved by a number of different super-peers, yet another super-peer is elected which maintains links to these ones. Thus, a tree-like overlay network is built over time, with super-peers as branches. By referring to super-peers, normal peers can locate provider compositions for their service requests if these requests have been resolved (or at least partly resolved) in the network. Specifically, a service discovery process will firstly examine super-peers and then probe the normal service providers if the requested service cannot be fully found in the super-peers. However, it assumes the network has super-peers that are capable of managing network links. This assumption is not safe in dynamic environments, as super-peer-based groups may fail if any relevant peer leaves the network, or is no longer able to fulfil its role because of reduced capacity. In addition, this work assumes offline planning, which limits the capability of the planning solution to cope with dynamic adaptabilities, such as dynamic re-planning.

2.3.4 Dynamic Controlled Flooding

Request routing solutions include (static) controlled flooding, distributed directories, and overlays. Hybrid routing approaches that integrate service caches/directories into an overlay network [24, 39] are most efficient in terms of network communications for service discovery. However, existing hybrid routing solutions relying on service advertisement [39] or the previous composition records [24] to maintain the discovery infrastructure. The former requires frequent service announcements for mobile service providers and overlay maintenance costs, and the latter may lead a composition request to an inaccurate destination. Directory-based solutions shown in Figure 2.7(a) and overlay-based solutions shown in Figure 2.7(d) improve search efficiency with short routing paths or less network traffic, but they also introduce maintenance cost of their support infrastructures. Optionally, the maintenance cost can be reduced by controlled flooding.

Controlled flooding-based routing shown in Figure 2.7(c) is infrastructure-less, and so has the potential to support highly dynamic environments. However, existing approaches rely on static flood controlling to route a request, which is inflexible and creates unnecessary communication. There is a need for a more flexible, infrastructure-less, dynamic controlled request routing to support service discovery in the target environment.

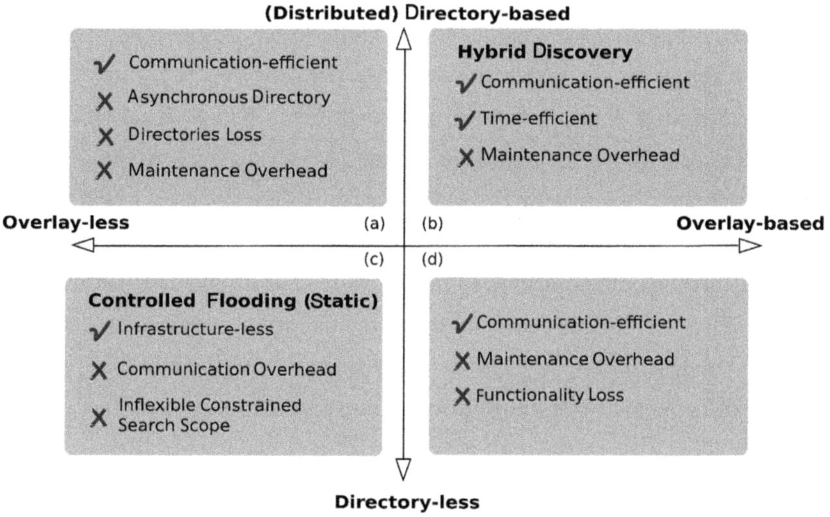

Figure 2.7   Analysis on composition request routing solutions
(a) distributed directory-based routing, (b) hybrid routing,
(c) static controlled routing, (d) overlay-based routing

Controlled flooding, as discussed in Section 2.3.4, is infrastructure-less, but current approaches either use static control mechanisms or require service providers to have a-prior topology knowledge about their local networks. Let us explore the novel

Dynamic Controlled Flooding as below.

Dynamic controlled flooding allows different edge devices in the network to independently decide how to route a composition request depending on a combined context, including their current local network properties, the cost that has been paid by other nodes for routing this request and the discovered service providers' properties. A request routing process can flexibly stop if a service provider is considered to be unreachable in a service execution process. Unreachable microservices are those that will introduce a high cost for routing execution results (i.e., more routing hops), which is in conflict with global execution time-constraints.

2.4   Composition Planning

Existing service composition techniques utilize dynamic composition planning mechanisms to reduce composition and execution failures while dealing with complex user requirements. Examples of such investigations applied to pervasive computing include open service discovery approaches and goal-driven planning approaches, both of which automatically discover a combination of multiple services to support a user goal or functionality when a single matched service is unavailable. They are differences in specifying composition plans. Composition planning can be specified by low-level requested service description④, high level requested service description⑤, and work-flow-provided service description[5]⑥. Open service discovery uses workflow-provided service description while goal-driven planning relies on high-level requested service description. They are flexible compared to those that rely on a predefined conceptual composite to find microservices that can exactly match its service requirement (i.e., low-level requested service description) [5]. This section discusses how flexible these planning approaches are, and how to self-organize the planning process.

2.4.1   Open Service Discovery

A graph-based service aggregation method [46] targets flexible planning in perva-

---

④ "... the requested service is specified as a workflow, given the set of atomic services to be composed".

⑤ "... the requested service is specified as a goal to be achieved".

⑥ "... it is assumed that service providers are able to specify workflows in which they can take part".

sive computing environments by modelling microservices and their I/O parameters in an aggregation graph based on parameter dependence among the microservices. Complex user requirements in this method are resolved according to predefined conceptual composites. The approach dynamically composes microservices to finalize an abstract microservice in a conceptual composite when no single microservice can support it independently. A composite service is found if the aggregation graph contains a path to link the abstract microservice's output parameter to its input parameter. Similarly, a dependency graph [31] was used for service aggregation, but its usage differs from that in the work of Wang Z et al [46].

since it directly maintains microservice dependency relations rather than their I/O parameters' relations. An open workflow [62] has been proposed to support service composition in mobile ad hoc networks. It models workflows that already exist in the environment as a super-graph, and discovers microservices through the data flow in the super-graph. Liu C Y et al. [26] proposed a parallel approach for service composition in pervasive environments that relies on function graphs to model complex composition requirements. This approach resolves a parallel function by simplifying its split-join logic, breaking the parallel branches. Specifically, it decomposes the corresponding parallel function graph as a set of sub-functions that require sequential service flows, and then searches for microservices to satisfy each sub-function. However, the above approaches require central entities or clusters⑦ for the graph-based service directory, which implies frequent network communications to maintain such a directory when the network topology changes quickly. In addition, they require a preexisting workflow to discover microservices that support complex data flows, such as one with parallel logic. Such a workflow may need to be generated offline by a domain expert or a composition planning engine, which is inconvenient when a change is required at runtime.

A decentralized reasoning system [63, 64], which does not need a pre-existing workflow, composes microservices using a distributed overlay network built over P2P networks and enables self-organizing service composition through management of the network. However, this approach assumes that all the participants know their geographic locations, and the service request is sent to the participants' physical neighbours. In mobile environments where devices' geographic locations can change quickly, frequently updating geographic locations wastes computation resources.

---

⑦ A cluster collects service information locally.

2.4.2　Goal-oriented Planning

Dynamic composition planning resolves user requirements and generates service flows during service composition. Classic AI planning algorithms, such as forward-chaining and backward-chaining, have been applied for dynamic composition planning. Web Service Planner (WSPR) [65] proposes a novel. AI planning-based algorithm for large-scale Web services. This work is based on an analysis of complex networks. The EQSQL-based planning algorithm [27] applies rank-based models to improve service composition efficiency. The work of Ukey et al. [47] models a Web service as a conversion from an input state to an output state and maintains published services as a dependency graph. It employs a bidirectional planning algorithm that combines a forward-chaining approach and a backward-chaining approach to find a path with the smallest cost from the dependency graph. Liu et al. [66] also proposed a bidirectional planning algorithm. They introduced tag-based semantics to specify composition goals, addressing effective service queries. Khakhkhar S et al. [25] improved bidirectional planning allowing systems to plan a composite service from the input data and the goal output at the same time. A planning solution emerges when the searches from the two directions meet at some point in the solution's service flow. However, these approaches require central service repositories to maintain service overlays, and they have no support for dynamic composition re-planning for composition failures. Web Service Modeling Ontology (WSMO) [19] describes single-direction planning models, which include forward or backward chain service providers for user tasks. WSMO reasons over service execution plans and adapts composite services on the fly, addressing flexible service composition, but still requires central controllers to schedule services.

PM4SWS [48] is a distributed framework to discover and compose web services. However, this service exploration process simply floods the network with query messages for service discovery, which is not suitable for pervasive environments, especially where there are large numbers of services to be considered [67, 68]. Geyik et al. [69] proposed a distributed implementation for their composition planning model. This approach uses a backward search scheme in a wireless sensor network to generate a service dependency graph that includes all the possible service paths from the services that can directly provide the requested output to every sensor node that does not require any input data. Each service provider in the graph only maintains links to its neighbouring services. After that, a composition graph can be generated by forwarding messages through the dependency links in the network. Although this approach supports distributed planning, it does not state that how a data-parallel task is resolved, and it invokes services only after all the potential service providers are discovered, which can delay the composition process and cause communication overhead. Similarly, a cooperative discovery model [24, 61] provides fully distributed support for unstructured P2P networks. This work also includes a bi-directional search model that allows a data-driven (finding services that match the query's input parameters) service query and a goal-driven service query that start concurrently to improve discovery efficiency. However, this bi-directional search scheme requires an initiator node to aggregate service information, which implies resource-rich devices.

2.4.3   Decentralized Flexible Backward Planning

Composition planning uses service availability information to reason about an executable service workflow. Generally, open and dynamic environments increase the possibility of diverse service flows being built for a single composition request. Goal-driven service planning is more flexible as it composes whatever usable services in the environment instead of using a pre-defined abstract composite to find only matched services. However, most goal-driven solutions still require a centralized planning engine and a service repository to support complex service workflows, which may plan faster, but any service information change during service execution is likely to make the primary planning result unreliable. Updating the service repository when changes occur may be a solution, but in a highly dynamic scenario, frequent updates on the service repository are expensive. Decentralized solutions tackle these problems by partitioning the planning process to allow local service providers or a set of brokers to collaboratively solve a composition goal. Unfortunately, existing decentralized goal-driven service composition approaches are not flexible enough to support parallel service flows.

A composition request reflects user requirements which may include data-parallel tasks, such as multi-source data aggregation. It is required to model user requirements in a flexible way to cope with various, diverse service networks. Generally, one way to enable service composition for such data-parallel tasks is to specify the tasks as abstract parallel workflows [23, 62]. An abstract parallel workflow includes control logic and ordered sub-tasks. Each of the sub-tasks can either match with either one basic microservice, or trigger a process to generate a composite service when there is no service that suffices independently. This way is inflexible at resolving data-parallel tasks. On the other hand, a composition process, in some cases, should be able to add new requirements into the original composition to satisfy data dependencies. Consider a navigation task as an example. The navigation has a service query including two requirements: a GetLocation service and a Navigator service that needs the results of a GetLocation service as input. In service composition, if a system can only find a Navigator service that uses both location data and map data as inputs, the system should be able to adapt the original service query list, by adding a requirement for getting map data. Under this kind of circumstances, a composition request is likely to be resolved by a parallel or hybrid service flow.

Assuming that no single node can maintain full knowledge of the network for execution planning, we resolve a data-parallel task through interactions between service providers, without prior knowledge of the task's inner data transaction or the global knowledge of available services. Moreover, this method adapts composition requirements, depending on available microservices to flexibly reason about a composition result that may be sequential, parallel, or hybrid. This is how **Decentralized Flexible Backward Planning** works:

It is inspired by sequential decision making of humans. Most of the time, humans make decisions on-demand instead of planning every step a-prior. This way ensures they always use the latest context information in the planning process and prevents them from information overloading, making it works well when dealing with complex and dynamic environments. Decentralized flexible backward planning relies on service providers receiving a composition request, and generating the potential fragments of the global execution plan. A composition request can be gradually solved through backward planning processes on each participating service provider, and a service provider can adjust a composition request when new data is required. This mechanism reasons about, if necessary, sequential, parallel, or hybrid service flows, using control logic, locally generating execution branches, and manages the service execution process.

2.5   Service Binding

A service composition model selects a group of microservices and binds their resources for invocation. To tackle dynamic environments, such a service selection could benefit from predictions for services' availability or service execution paths' strength (reliability). It is also worth considering microservices' Quality of Service (QoS) attributes. On-demand binding mechanisms are investigated here to address dynamism in another aspect, that is, to keep service selection flexible until a microservice has to be selected for immediate use. The analysis in this section includes consideration of how to select service providers and when to lock these service providers' resources to maximize the possibility of successful service invocation, while not impacting the possibility for candidate providers to be selected by other compositions. Note that approaches that assume random selection at an early stage of composition [41, 42, 70], and those assumed providers following a schedule to act [45] are not considered in this assessment.

2.5.1 QoS-based Selection

The work of Mokhtar et al. [71] takes non-functional properties of services into consideration when selecting service providers in pervasive computing environments. In this work, service providers predict their own QoS before service execution and enclose this QoS information into their service advertisement messages. The approach uses a set of QoS metrics to quantify QoS information, which includes a probability value for service availability and a value that indicates execution latency. This work uses probability values to define service availability, which only caters to when a service is withdrawn by providers, and not when the provider moves out of range. The work of De Medeiros et al. [72] predicts the quality of the entire service composition instead of the individual service provider. It models service compositions' cost and reliability, defining six cost behaviours and four workflow reduction models for different service flows. However, they do not consider the reliability and the cost of communication among successive services.

Hierarchical Service Composition Framework in Service Overlay Network (HOSSON) [73] maintains a user Perspective Service Overlay Network (PSON) for all the candidate service providers. The PSON is graph-based and performs service selection using multiple criteria decision making to find the optimal service execution path with minimal service execution cost and good reputation⑧. The work of Wang et al. [46]

---

⑧ Reputation is a ranking value that is given by previous service users to represent a service's dependability.

also considers the quality of overall service execution paths when selecting service providers. It adopts path measurement including reliability, execution latency, and network conditions. However, the above approaches assume the QoS information is predefined by service providers, and as network conditions are likely to change, this assumption is not safe in a dynamic environment.

Zhou et al. [74] selected services at an early stage in a service discovery process, depending on the strength of service links. During service advertisement, if a service provider announces its service specification, a neighbouring node caches the service specification and also maintains information relating to the strength of the path to the service provider. When a node receives a service request, assuming information about multiple functionally equivalent service providers that match the request is cached, the node selects the most reliable path to itself. When a selected destination is unreachable, a backtracking mechanism can be performed to recover service binding. Though path strengths are involved in binding decision making, and a binding recovery mechanism is proposed for failed routing, this approach is still unsafe as beforehand cached path strength information may be out of date, and the backtracking-based recovery is expensive itself. Surrogate Models are used to facilitate service composition in mobile ad hoc networks [75].

2.5.2 Adaptable Binding

ProAdapt [76] monitors operating environments and adapts service composites to a list of context changes, including response time changes, availability of services, and availability of service providers. Fault Tolerant Service Selection Framework (FTSSF) [77] applies a monitoring and fault handling process for service provisioning in pervasive computing. Monitoring is concurrent with a service delivery process, allowing for service re-selection if the execution crashes. The work of Prinz et al. [21] allocates a service provider to the best performing candidate in terms of the QoS while keeping a group of backup providers. A backup provider can replace a previously bound provider if its execution fails. This model is more efficient than ProAdapt and FTSSF, as it monitors the service composite only when it is executing, which means there is no need for a service composition system to continuously monitor the operating system throughout the composition process.

Wang et al. [22] modeled service composition as a problem of finding a service provider for each abstract service in a predefined conceptual composite. Service binding adaptation in this approach is based on automatic QoS prediction. Specifically, a service composition system firstly allocates a set of services to form an execution path, the QoS of which conforms to the QoS constraints that are defined in the composition request. The system then predicts the failure probability of the path, and finalizes the service composite according to the prediction result. If necessary, re-selection for providers can be performed. This solution regards the service execution path's reliability as an important criterion for service selection and allocation instead of considering each of the service providers independently. Ravindra et al. [78] further considered edge computing environments. They adapted the data transmission service to the availability and computing capacity of edge computing devices to ensure an acceptable QoS. However, the above approach relies on a centralized composition handler to perform their QoS prediction.

2.5.3   On-demand Binding

Open Service Infrastructure for Reliable and Integrated Process Support (OSIRIS) [158] selects service providers on-demand at execution time. The selection of service providers depends on runtime device load and dynamic invocation cost. Service discovery is performed offline to find candidate service providers but repeated to detect new service providers. However, the approach depends on a central repository to store service specifications, which is limited in dynamic systems.

Opportunistic service composition [23, 79, 80] also proposes an on-demand service binding mechanism. Unlike OSIRIS, opportunistic service composition does not use central repositories to keep service specifications. Instead, it discovers service providers on the fly relying on request flooding. After service providers are located, the approach asks for permissions to lock service providers' resources, and then invokes the service. This model ensures an available service provider will be invoked for execution, but refinements are required to further consider result routing when binding a service provider, at the same time reducing the cost of its flooding-based service discovery and increasing the flexibility of composition planning.

2.5.4   Path Reliability-driven Selection

QoS-based service binding addresses dynamic environments by self-describing services' runtime properties (e.g., availability, reliability, response time, etc.). Existing approaches [46, 71, 73, 74] depend on QoS descriptions provided by service providers that predict their own service performance. Although this mechanism is lightweight for a service composition system as no monitoring effort is required, such QoS descriptions are likely to be inaccurate. Moreover, mobile service providers may have to frequently update their QoS description. Adaptable binding detects changes in operating environments and adapts a service composite accordingly. Detecting changes or failures requires different levels of monitoring. Execution time monitoring is the most efficient, but failures can only be detected after they occur. Recovering a composition from emerged failures can introduce additional time and communication costs. On-demand binding selects service providers using up to date service information, requiring no extra environment monitoring or infrastructure maintaining efforts. It would be interesting to improve the existing on-demand service binding mechanisms to support reliable service composition, and at the same time, allowing the bound service to smoothly and seamlessly adapt to context changes.

To reduce execution latency, a composition process selects service providers that can form the shortest service flow. Moreover, the execution time of each service should be considered as a factor of execution latency. Routes between successive service providers are also required to be short, which reduces the possibility of failures when invoking service providers or routing intermediate execution results. This is how **Path Reliability-driven Selection** works:

It selects services based on the path's robustness (reliability). A path's robustness value is calculated using multiple criteria including the execution time of each individual service, the length of service flows, and the routing hops during service discovery, which indicates whether a remaining execution path is likely to be reliable in a period of time.

2.5.5 Bind Microservices on-demand

On-demand service binding is flexible, reducing standby time for service providers. Current on-demand binding approaches [23] bind multiple functionally equal service providers to one composition process during service discovery and send a token message to release them after a service provider is selected for invocation. Although the duration between binding and releasing may be small, many of the service providers that support the required functionality may become unavailable for other composition processes in the network during the time of standby. On the other hand, releasing service providers introduces traffic overheads. Thus, this book introduces the method of **Bind Microservices on-demand**:

We propose a dynamic composition overlay that organizes service providers that are currently participating in composition processes. The dynamic composition overlay is temporary and only exists when a composition process is performing in the environment. Using this overlay network, a service provider is bound only when it has to be invoked.

2.6   Service Invocation

This section analyzes how service execution processes manage mutable environments. Centralized service invocation has been studied in many solutions [71, 81-83], however, supervising service providers' execution with a central entity limits mobile-awareness and consequently is less flexible. Research on distributed service execution includes the following solutions: Fragments distribution approaches partition a service composite and distribute each of the abstract microservices in the composite as a task for selected service provider, service composition is managed in a decentralized way by these selected entities. Process migration approaches only partition the service execution process, and issue a full composite to the first provider (or a primal networked broker), the composite is partially resolved by the first provider and the rest of the composite is handed over to the subsequent providers (or backup networked brokers). In this section, the analysis of service invocation focuses on the question of how a service composition model self-organizes the composition process.

2.6.1   Fragments Distribution

The work of Prinz et al. [21] distributes the execution process to a set of selected service providers and maintains references to the other candidate providers as backups. The execution of selected providers is monitored by backup providers. Once a selected provider fails, one of the backup providers takes over the execution process. This approach achieves flexible decentralized service execution, but its service discovery and selection process still require a central composition engine.

Fdhila et al. [84] targeted service compositions that contain complex control logic, which is decomposed using a dependency table. In particular, the composition model slices a conceptual service composite into a set of subprocesses, each of which is realized by only one service instance. These sub-processes are allocated to associated service providers and executed independently. However, this model assumes a global knowledge of available service providers, which is difficult in a dynamic environment. OSIRIS [32] decomposes a central conceptual composite description into a set of execution fragments and distributes these fragments to different service providers. It also supports complex control logic but assumes a continuously available node to manage a parallel service execution.

2.6.2 Process Migration Approaches

Chakraborty et al. [85] proposed distributed broker-based model for service composition. The approach dynamically selects a group of networked entities, which relay a composition process from one to another. A requester initially assigns a composition task to one broker. This broker performs service composition until, potentially, a failure occurs. If the composition stops at this broker because of a failure, its current state, and any intermediate execution results are frozen and handed over to another broker that executes the remaining process. The new broker's address is sent to the requester. Selecting brokers and creating a network with the selected brokers are key to this approach. Anything like brokers' mobility or reliability can affect the service invocation process. However, it is difficult to create a good broker network in the target environment where a global view or a central controller is impossible.

A decentralized Workflow Management System (WFMS) [86] that self-describes a workflow and relays the workflow from one service provider to its subsequent service providers, resolving the workflow partially on each participating provider. Similarly, continuation-passing messaging [87] is a decentralized execution model that passes a service execution process from the primal service provider to the rest of the providers. This model selects a node in the network to handle execution faults, named the scope manager. When a service provider detects a failure, it sends the failure information to the scope manager. The scope manager receives the failure information and re-executes services. Service execution in this model is fully decentralized, however, failure recovery is limited by the resource and communication range of the scope manager.

A process instance migration approach [88] allocates the full control logic of a conceptual composite to every participating service provider. It allows participants to keep composition information so that each participant can select a granularity of fragmentation at runtime. The approach is required to be further improved to support a decentralized composition adaptation model.

2.6.3   Runtime Service Announcement

Process migration allows service providers to know a part of a request and gives them the chance to alter the part of the process they know by using their knowledge about the local networks. Fragments distribution decouples the execution process into execution fragments and allocates them to corresponding service providers. Fragments distribution minimizes process duplication but requires a central entity to be responsible for service discovery and allocation. Both of the methods resolve service invocation in a decentralized manner. To achieve successful execution and protect runtime data flow, decentralized service invocation needs to efficiently adapt to environmental changes, and prevent the composition process from exposing the full data flow. In terms of dynamic adaptation, process migration is more flexible than the fragments distribution approaches when supporting functional service replacement. This means when functionality is no longer required before its matched services get invoked, a process migration manager will have the potential to remove the functionality from the composition before it hands over the process to the subsequent manager, while a service provider in a fragments distribution approach can only replace a failed service by another that supports the same functionality. On the other hand, migrating a composition process from one service provider to another may reveal the composition's control logic and data flow to those providers. Generally, process migration is more flexible to realize dynamic adaptation, and fragments distribution can protect the overall data flow from being observed by a single third party.

Most service composition approaches that assume periodic service announcements are expansive because of frequent service matchmaking and multi-hop broadcasting. Normally, service announcements are used to collect information about locally available services and create a directory or overlay structure to maintain such information for service discovery. If a new service provider joins a network by announcing its microservices, all the neighbouring nodes of the service provider have to analyze the announced service specification (service matchmaking). Service announcement usually relies on multi-hop broadcasting messages that contain service specifications, but multi-hop broadcasting is expensive itself[89, 90]. Runtime Service Announcement may be a usable trade-off:

This book distinguishes service providers, allowing the service provider that is currently participating in a composition process to analyze a new service and decide whether to invite its provider to join the composition. The service providers that are not engaging in any composition process (idle providers) use announcement messages only to obtain a sense of local network properties, such as service density, to get prepared for future compositions. We allow service providers to announce their services to the network, using one-hop broadcasting. The idle providers receive service announcement messages, instead of calling a matchmaker to launch an expansive semantic service matching process [95] [9], they compute only network properties.

2.7   Fault Tolerance

Fault tolerance mechanisms can be classified in terms of how to deal with the potential failure and actual failure throughout the composition process. Many approaches have explored fault tolerance performing after an actual failure is detected, such as Forward Central Dynamic and Available Approach (FCDAA) [91] and the RVNS approach [53]. Preventive adaptation [44, 69, 92, 93] focuses on anticipating potential failures and taking actions to prevent them. Composition recovery should have a fast, communication-efficient model to recover a faulty composition, without disruption. This section discusses how to detect and handle failures in a timely fashion, and if the additional cost for fault tolerance is affordable for mobile and pervasive environments.

2.7.1   Preventive Adaptation

The OSIRIS approach [32] decomposes a central service flow description into a set of execution units that can be deployed on service providers in a P2P network. These service providers are found during service discovery. The service execution process migrates from one service provider to another. Each time such migration occurs, the client node can select another available node. It defines special observer nodes to monitor the nodes that may cause failure. If a failed node is detected, the execution instance can be migrated to another available node. However, it starts execution after service discovery is finished, and it requires that a part of the execution unit is de-

_____

[9] Semantic matchmaking takes 0.08-10.66 s to return a result depending on matchmakers.

ployed to all the providers. The approach provides no means to allow a potential new provider that may appear in the network to participate in the composition. Multi-channel Adaptive Information Systems (MAIS) [94] is an adaptive service composition model for web services. It reduces service invocation failures by negotiating QoS in advance of composition selection and execution. Although MAIS provides strategies to decrease negotiation overhead at runtime, its negotiation uses previously cached information about services and QoS which is likely to be changed in a dynamic environment.

A fuzzy-based service composition approach [29] has been introduced for mobile ad hoc networks, which considers resource-constrained devices and error-prone wireless communication channels. Each node maintains its neighbours' service information and gets real-time QoS information during service discovery. A fuzzy Technique for Order of Preference by Similarity to Ideal Solution (TOPSIS) method [96], unlike the approach of Prochart et al. [29] that has no support for adapting composite services at runtime, adapts service bindings according to real-time user preferences. TOPSIS can prevent unnecessary adaptation through fuzzy analysis, but it has no support for adaptation triggered by operating environment changes. A distributed dynamic composition model [69] allows a service provider participating in the composition graph to notify its successive services' providers when it changes its own service information and raises the potential of adapting a service composite before an invocation failure occurs. This model states a way to find an alternative service from a distributed service dependency graph. However, it is not clear that how this can be achieved during the service composite's execution.

In the era of edge computing, time-efficient services have become more popular, and preventive adaptation is widely used to tackle the resource constraint environment and increase resource utilization. The AFC approach [44] is proposed for battery-powered service providers and introduces an optimization and sampling method to adapt fog configuration to service demand. Similarly, Lin et al. [92] provided an adaptable fog configuration method to redeploy data services to reduce delay on service access. To prevent QoS decreasing, the work of Zhang et al. [93] dynamically adjusts the service rate according to the network constraints and the service consumer's QoS requirements. However, the current solutions mainly focus on simple services such as data transmission or task offloading. It remains a requirement to deal with timely fault tolerance for applications with complex functions or flexible workflows.

2.7.2 Composition Recovery

The work of Yu[87] uses acknowledge messages to monitor service execution. It also allows a service execution process to retry a service invocation after it fails. Although this policy targets unreliable wireless channels, giving a service provider a second chance to offer its services, it has limited and inflexible support for composition recovery. The work of Prinz [21] provides a recovery policy that is more flexible than the one proposed by [87]. Instead of retrying failed invocations, [21] keeps candidate service providers during service execution, allowing them to monitor the primary service provider's execution. When the primary one fails, one of the candidate service providers takes over the service provisioning process. However, this policy requires primary service providers to push heartbeat messages to backup providers to indicate its execution status. If a service's execution has high latency, pushing heartbeat messages can quickly exhaust its providers' battery, and in turn, reduce the availability of the service.

TLPlan [82] allows a system to plan an abstract solution from a pre-existing abstract service repository, and then to discover and bind services for execution. In each step of the service composition, TLPlan provides on the fly rollback mechanisms to deal with potential faults like composition failures, service discovery failures, or service execution failures. TLPlan, however, plans for a composition offline, relies on central composition engines that have not yet been applied on mobile devices and re-generates a new plan when a composition recovery is required, which is time-consuming and not suitable for dynamic environments.

A minimum disruption service composition model [97] investigates the types of composition failures and provides network-level and service-level adaptation, as well as recovery mechanisms. During service execution, the approach quantitatively estimates the one-hop forward service execution path's lifetime according to the distance to the next service provider and composes a new service path if the next service provider is missing.

A cache-based service execution and recovery model [74] caches backup services. The approach adopts a notion of a magnetic field to underpin an adaptation policy for cached service information in the vicinity. A service provider initializes a magnetic strength value that determines the border of the vicinity, which indicates the maximum transmission hops for its advertisement and counts down by one for each message relay hop. A service provider periodically advertises its services, and its physical neighbours cache the provider's functionality and a magnetic strength value. The closest neighbour will rank this service provider the highest. During service discovery, if a service request reaches one of its neighbours, the neighbour can route the request to the service provider, and if there are multiple service providers cached by the neighbour, according to their magnetic strength values, the service provider with the minimal transmission hops is selected for request routing. A backtracking strategy is applied when a selected service provider has moved away before the request has been routed to it. A failure message is routed backward via the routing mediators towards the request source. If one of the routing mediators has cached another service provider that supports the same functionality as the missing one, the new service provider is selected and the request is re-routed from this routing mediator. This backtracking strategy allows for a short recovery path and less communication overhead, but it requires frequent updates to the cached service provider information if the network topology changes quickly, and it is unclear how the backtracking works if a routing mediator leaves the network.

2.7.3   Local Execution Path Maintenance

Preventive adaptation is timely as a service execution will not be interrupted by failures and recovering processes. However, existing solutions rely on monitoring infrastructures or central decision-makers to determine adaptation actions, which is infeasible in our target environment. Composition recovery aims to explore an efficient model to recover a faulted composition. Current solution includes keeping backup service providers [21, 74] for quick replacement, dynamic path re-generation [97], and rollback composition [82]. Using backup service providers is efficient but inflexible because the backup service providers are previously found and may become unavailable at runtime. Rollback composition re-plans for composition, which is flexible. However, re-planning on the fly is likely to be time-consuming and delays the composition result. A new fault tolerance model is needed to dynamically re-planning for (a part of) execution path, without affecting currently executing services.

An execution path for a composition may need to be adapted to the environment. Global knowledge of available services is infeasible in our target environment, and so a composition adaptation process should be performed locally, without affecting the global QoS of the composition. **Local Execution Path Maintenance** works in such a manner:

We create a fragment of an execution path on individual service provider. Path fragments are maintained locally by merging new service providers or removing the parts of a path fragment that become invalid.

2.8 Chapter Summary

This chapter proposes a set of design concepts and reviews the state of the art of service-oriented computing model in pervasive computing from the perspective of openness and dynamism. The most related solutions include efficient service discovery [30, 39], flexible planning [20, 24, 31, 35], dynamic binding [23, 79, 80], decentralized composition management [21, 23, 24, 45] and dynamic fault tolerance [19, 21, 74, 98, 99].

The proposed design concepts target the following open gaps:

Inflexible and heavyweight to compose microservices in dynamic pervasive environments. Mobile service providers need to self-organize a service composition process that can flexibly use local service knowledge to plan for a global composition result. The process of composing relevant services is itself likely to be time-consuming, and so models are required to streamline or otherwise ensure that the service composition can be done within time constraints.

The limited support of adaptation for service composites. An appropriate granularity of adaptation model is essential for a service provisioning model to quickly adapt a service composite while without introducing a heavyweight adaptation model.

Chapter 3

Microservice Deployment in Edge/Fog Computing Environments

This chapter introduces a fog as a service model [100] and a network framework that support fog as a service that enables collaboration among edge devices to ensure acceptable end-to-end application performance in dynamic computing environments. It enables microservice deployment for pervasive applications. This model pools resources and enables an application management overlay network that spans along the cloud-to-things continuum including in the cloud, at the network perimeter, or on the things. It means microservices can be deployed anywhere along such a continuum. This chapter will further explore the service provisioning challenges and the edge computing environments' features, introduce the architecture, the service management, and the deployment mechanism of the fog as a service model. It will also investigate the challenges regarding service adaptation given the dynamic computing environments.

3.1   Edge Computing: Pervasive Applications' New Enabler

The comprehensive development of IoT, AI, and 5G will drive more intelligent pervasive applications, and also pose new challenges to the underlying computing and communication system architecture. IoT will support substantially more end users and devices. It encompasses massively diverse entities such as vertical systems, communication networks, smart devices/things, and applications. These entities enrich IoT with a plethora of data [101]. Such data and the great amount of new connected things will trigger increasing demands for new pervasive applications.

The pervasive applications make use of such data and provide timely actions or feedback to the IoT users. An IoT system needs to collect, process, and analyze data from many sources in a timely manner and provide services not only to users in a special vertical domain but also to those have cross-domain functionality requirements [101].

The current solutions for pervasive application development over IoT environments generally rely on integrated service-oriented programming platforms. In particular, resources (e.g., sensory data, computing resource, and control information) are modeled as services and deployed in the cloud or at the edge [12]. Cloud computing offers ubiquitous, convenient, and on-demand computing services by allowing resource constrained nodes of a network (e.g., devices) to access a shared pool of computing and storage resources [102]. Software as a Service (SaaS), Platform as a Service (PaaS), and Infrastructure as a Service (IaaS), are the main cloud computing services offered to resource-constrained nodes.

Cloud servers are generally used for centralized control and are responsible for handling global tasks. Due to the long distance of cloud servers from IoT nodes and limited network bandwidth, it is difficult to provide real-time data processing services. Although such services are advantageous in several ways, they have multiple limitations when they are required to satisfy requirements such as low latency, context awareness, mobility support, data safety, and privacy [103].

Recent advances in mobile computing and wireless communications bring new opportunities and challenges to pervasive application development. Many IoT-based applications, such as smart cities where huge numbers of IoT devices are used by smart city services and applications, are shifting their paradigm from cloud computing towards edge computing [11, 104]. Many edge devices offer a variety of resources and services including local computing, storage, and security services. These devices can act as smart gateways to interconnect other devices in the surrounding areas and share their resources with each other and with the end users. For example, in intelligent transportation systems, self-driving vehicles can benefit from the low latency communication provided by mobile edge computing to coordinate their behaviours and make decisions in real-time [105]; applications such as video [106] and online games [107] can also use edge services to obtain efficient computing and communication services, thus improving the quality of experience (QoE) of users. When dealing with data explosion and network traffic, the need for smarter computing paradigms and sustainable energy consumption is the key reasons why many applications choose edge computing [108]. With edge computing, pervasive applications such as smart cities will come closer to reality by employing different electronic devices, sensors, and actuators in various applications to improve the QoS in systems such as smart homes, surveillance systems, environmental protection, vehicular traffic and transportation, healthcare, and weather and water management systems [109].

Edge computing was originally proposed jointly by IBM and AKAMAI, and is offered on their joint platform, WebSphere, as an edge-based computing service. Edge computing is a technology that provides services close to the data source. That is, the edge computing model shifts control of network services from a central node (such as the cloud) to the end of the network (a.k.a., the edge). Edge computing deploys computing resources near end devices and uses those resources for local storage and initial data processing to alleviate network congestion while increasing the speed of data processing and analysis, thereby reducing latency.

In the edge computing family, there are many different members that shares very similar concepts, including Cloudlet, fog computing, mobile edge computing, multi-access edge computing, multi-tier computing, etc. Introduction of the more popular 3 ones among these is as below.

Cloudlet, introduced by Carnegie Mellon University, is a computing model that shares the same technical standards as cloud, but its computation resource is closer to the user than cloud. Cloudlet can provide the same services as cloud computing, but with limited resources, unlike cloud computing, which can provide nearly unlimited resources.

Fog Computing was first introduced at Cisco Live 2014, which emphasized that fog computing is a new computing model based on today's ubiquitous IoT applications. Fog computing changes the way services are delivered to customers to meet IoT requirements and targets the challenges in the pervasive environments. Fog infrastructure and service platform will not only be at the network perimeter but also span along the cloud-to-things continuum to allow computing resources to be deployed anywhere along this continuum, including in the cloud, at the edge or on the things, and to also pool these distributed resources to support applications [12]. It has great potential to off-load tasks from the cloud to fog service providers that reside in the vicinity of the end users or the data sources, which will reduce the latency and bandwidth required for transporting data to the cloud. Multi-tier networks are extensions of fog computing, which aim to build a network infrastructure with hierarchical computing resources to process data from multiple data sources.

Mobile Edge Computing allows base stations in mobile cellular networks to serve as service delivery nodes, and multi-access edge computing is extended by mobile edge computing, which allows multiple network access methods when an end device accesses the base station.

Since the above techniques share a common goal that is to reduce end-to-end latency and network congestion, this book does not emphasize the difference among these techniques when mention edge computing.

3.2 Features in Edge Computing Environments

Most end devices have constrained computing resources due to their battery-charged power model, limited size, and low-cost design purposes. Edge computing brings computing close to the end devices (a.k.a., data source) and provides computational processing, storage, caching, and other services for data users [110, 111]. Edge devices, as service providers, should be designed to meet requirements such as low latency, context-awareness, and mobility support. The service provisioning process must be reliable and scalable [9].

3.2.1   Latency-sensitive

Many pervasive applications are time-sensitive and therefore require data processing directly at the network edge [112-114]. In the same application domain, varying services' latency requirements can be different. For example, in an in-vehicle application of self-driving cars, navigators and obstacle detectors work together. The navigator service has a milliseconds-to-seconds level of tolerance for latency. As vehicles move quickly, and avoidance of obstacles is directly related to human safety, the latency of obstacle detection must be at a milliseconds level or even smaller. This requires the edge device to be able to adjust the priority of various computing tasks and resource allocation according to the QoS needs. For example, when there are both navigation and obstacle detection demands, the task of obstacle detection can have higher priority to access nearer services or be allocated richer networking and computation resources.

3.2.2   Mobility is Everywhere

With the rapid development of wireless communication technology, an increasing number of end devices communicate with edge servers through cellular networks or WiFi, and the network topology between mobile devices and edge servers keeps changing [115]. It happens that a device was at one location at a previous moment and requested an edge device in its vicinity to perform a computation task and then moved to another location when the edge device finished the computation and became ready to send back the execution results. This requires the edge network to be able to track the change of the device's location and route the computation results back to the device on time [116]. Also, mobile devices in the network need to be able to search the network environment and update the list of neighbouring edge devices after their locations have changed.

Services of interests rely on the collaboration and resource coordination of multiple edge devices and end devices that may move dynamically in the environment. In such uncertain and highly dynamic environments, edge computing must be flexible enough to handle these changes and manage resources to ensure service availability in a self-organized way.

3.2.3 Openness of Network Systems

In an edge computing network, new devices can join or leave the network because of their mobility. It will also happen that some devices withdraw services due to failures, energy shortages, or other reasons. These facts lead to a constantly changing network topology [116, 117].

With new kinds of end devices invented and put into use with the development of technology, data sources can also change. For example, traditional cashier services only supported cash collection, but later were able to support POS (Point of Sales) machine for card payments, QR (Quick Response) code for cell phone payments, and even the payment methods based on biometrics technologies such as face recognition, fingerprint or palm-print recognition, etc.

Therefore, services that achieve the same function may also involve several different types of end devices, terminal services, or data sources [93, 115]. Service provisioning over edge computing networks should be able to deal with the openness of the system and continuously adjust to changes in type and size of devices in the network topology, data source, application requirements, etc.

3.2.4 Constantly Changing Environment

End devices work in real physical environments and are affected by a variety of external environmental factors, such as wind, rain, noise, interference, light, temperature, obstacles, etc. [118]. The changes in the external environment are not controlled by the software system, so it is not possible to perform operations such as scheduling and tuning of the environment. There exist ways to monitor changes in the environment but it is difficult to handle them in a timely manner at the end device level due to the limited capabilities of end devices [119]. Unpredictable events such as device failure or network unavailability will also affect the QoS of the entire system.

Edge computing systems are required to be able to flexibly adjust its own architecture as well as configurations according to the changes in the environment. In addition, the wireless channels on which cellular or WiFi communication depends are also unstable, so the transmission bandwidth of the network can change as well. The system has to be able to adjust itself to the changes in network bandwidth during service provisioning [92].

3.2.5 Limited Power Supply

Many edge devices are powered by batteries, which makes them reserve a certain amount of power for possible upcoming tasks before running out of energy [38]. In recent years, a part of edge devices have started to use renewable energy as power sources [118], such as wind and solar energy. Compared with traditional power supply methods such as thermal-based power production, renewable energy sources are widely distributed and with less or neglectable polluting to the environment, but the power supply is often unstable and easily affected by the natural environment. Therefore, when the service provisioning system includes edge devices rely on renewable energy, it is necessary to always monitor the changes in the natural environment and predict the possible power charging opportunity in the near future [38]. Such limited power supply situations require the overall service provisioning system to consider the past experience, the current task status and the current and future energy situation to reasonably allocate tasks.

3.3 Fog as a Service Model

Faster data transmission speed, higher network capacity, and wider bandwidth, on the other side, have the potential to support more complex data transactions between devices [120]. Therefore, more flexible and efficient applications become possible by employing edge devices as service providers or application controllers. Cross-domain applications that require services from different domains will become a popular kind of pervasive applications. For example, an emergency response request may trigger services from healthcare, transportation, traffic, etc. Currently, many edge devices and edge computing networks are deployed by application providers. It creates application silos at the network edge or on the things, which limit the potential of those services to be shared by other applications. Specifically, developers have to develop and deploy applications for many different edge devices and things, resulting in heavyweight deployment effort and service redundancy.

To bring such opportunities to reality, a service enabling model should target the low resource availability issue and the seamless service provisioning requirements as well as support for cross-domain applications.

Improve Resource Availability

IoT systems must improve their resource availability to cope with frequent resource access, especially when end users and application demands are increasing. As Figure 3.1 shows, engaging smarter and capable edge devices that reside between cloud and things in an IoT application can greatly increase available resources and services, but there are challenges to manage different granularities in devices and communication channels, and to use these dynamic resources efficiently.

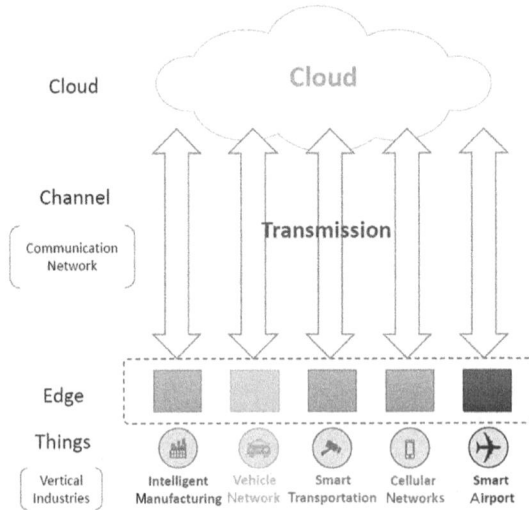

Figure 3.1   The cloud-edge-things system [100]

Seamless Service Provisioning in Dynamic Networks

Mobile users often require localized services. To meet such demands, it is essential to provide services in a user's vicinity since hauling local information to and retrieving information from remote clouds tend to be inefficient. This calls for IoT applications to access resources without knowing the resources' physical locations or network locations while the end user is moving from one location to another [121].

Support Cross-Domain Applications

IoT environments require closer integration of applications from different application verticals. But heterogeneous devices, networks, and data semantics complicate interoperability and turn the collaboration across multiple vertical systems into a challenging, heavyweight and time-consuming task [122].

Fog as a service is a business model that allows diverse service providers to deploy and operate computing, storage, and control services at different scales, regardless of whether they are large organizations, small companies, or even an individual person. Figure 3.2 illustrates where fog as a service works. Fog as a service requires innovative technology support in the infrastructure, platform, software, and service levels. Such technology supports will achieve the full potential of edge computing and boom the boost of novel pervasive applications.

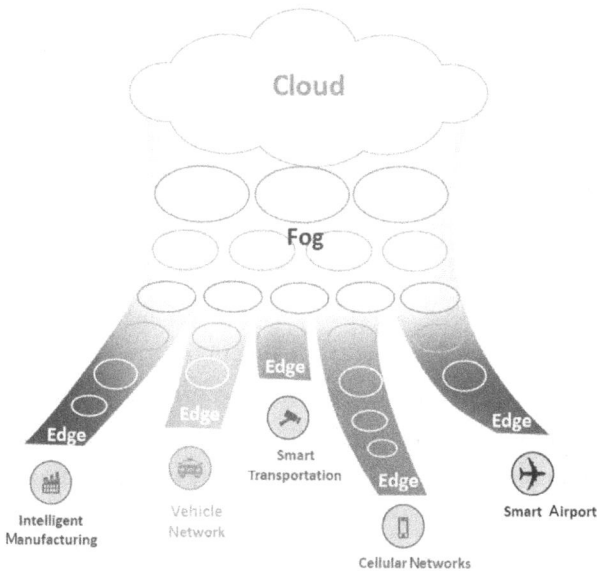

Figure 3.2 A fog as a service system usually works between cloud and edge [100]

The existing solutions, including mashup service [123], AWS Greengrass [121], fog orchestration [115], and microservices [1], all seek to provide services close to the end users [120, 124].

To underpin service provisioning in dynamic environments, this chapter introduces fog as a service technology and its architecture. This architecture aims to enable a multi-level service management system for fog as a service. It supports a cloud-to-things service continuum from service providers' perspective by a fog choreography system that concordances different communication channels, storage and computation resources to provide and manage microservices for varying application domains. The core of this architecture is a dynamic fog network that manages service deployment and localizes resources for an efficient transition process, and an adaptable service management model to wrap fog resources, also to enable fine-grained service provisioning across multiple application domains.

3.4 Edge/Fog Computing Architecture

Consider an IoT environment that supports multiple applications in different vertical domains. Three applications are deployed in the environment: intelligent manufacturing, smart transportation, and cellular networks, each of which is formed by a composite of atomic services. An atomic service is a minimal entity that provides software functions and is defined by unified interfaces for service access[115]. Atomic services are referred to as microservices in this book for simplicity. Figure 3.3 shows architectures of the current cloud-based and fog-based applications in this case[125]. Note that, in many cases, the term microservice is used to differentiate from monolith services and primarily denotes a distributed service-oriented architecture that uses small and concentrate services.

Relying on technologies like network function virtualization, software-defined networking, and data centers, a microservice can provide functions like computing, networking, controlling, or data storage. Such microservices can be deployed in cloud servers or any device that supports remote service invocation. In this book, we model this kind of device as fog nodes, extend our previous solution for resource-rich fog nodes[126], and realize a lightweight fog-based service provisioning architecture[125]. As illustrated in Figure 3.4, this architecture consists of two layers named fog infrastructure and fog-based service platform to meet user requirements on resource availability, architectural scalability, seamless service provisioning and cross-domain application development. The fog infrastructure can manage different types of devices, maintain their network connectivity, as well as visualize the computing, networking, and storage resources. It allows smart devices to participate in service enablement as fog nodes and share their resources to their computing fog infrastructure and is invisible to end users. It relies on IoT middlewares, operating systems, and node management modules to build a horizontal computing platform across different systems, devices, and networks to allow software vendors to develop IoT applications over a distributed fog infrastructure.

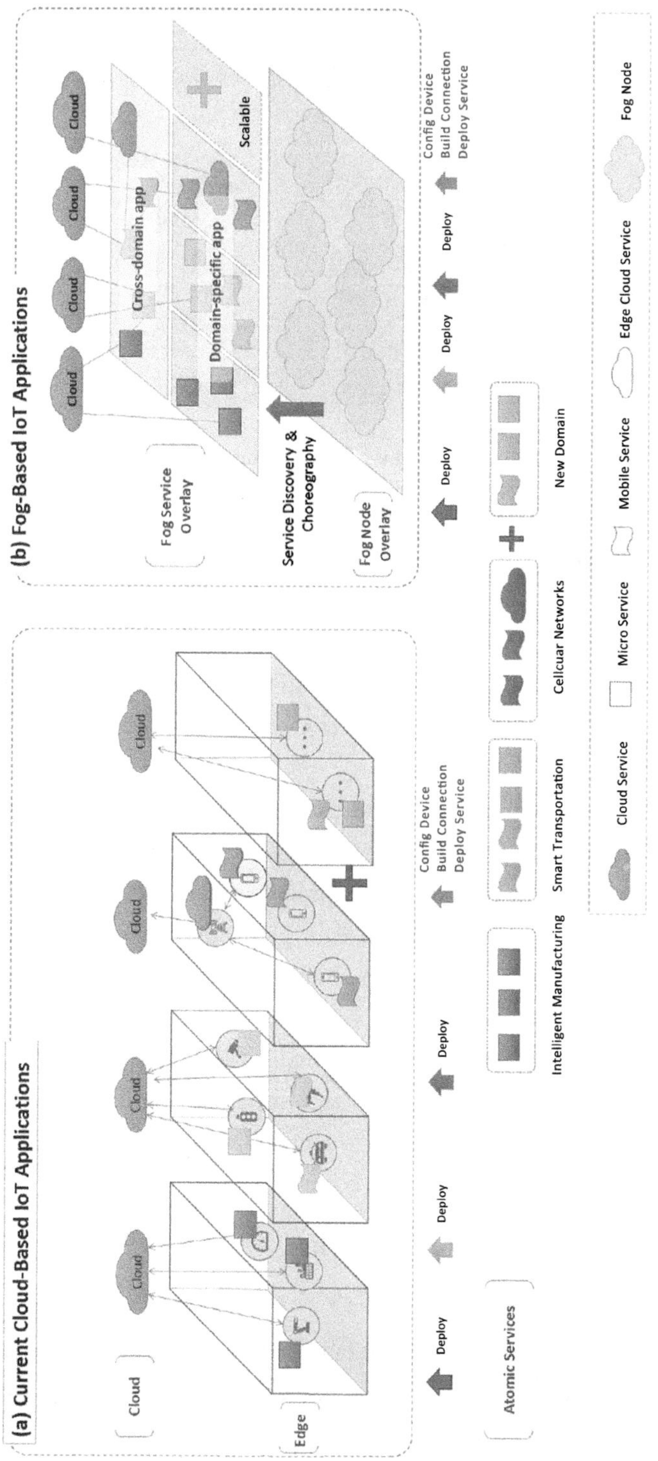

Figure 3.3   Architectures of current cloud-based IoT applications and fog-based IoT applications

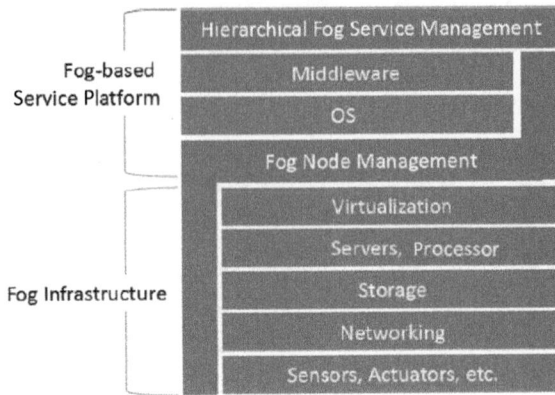

Figure 3.4　Fog-based service enabling architecture [125]

The core of this platform is a hierarchical fog service management module and a fog node management module. The former provides the application's functionality and behaviour to the fog environment as microservices and supports runtime service-to-service collaboration to ensure the execution of applications, and the latter is a cross-cutting module which manages collaborative fog nodes on a Fog Node Overlay Network (FON) for service discovery and ensures appropriate software and hardware configurations for the service execution.

In an application, the fog infrastructure organizes fog nodes and opens their computing, networking, sensing and storage resources to the computing environment. Software vendors implement and deploy microservices on appropriate fog nodes through the fog-based service platform. The platform manages the computing and storage resources for those microservices, hiding an application's implementation details and providing microservice API (e.g., service descriptions) to make the microservices accessible. Microservices are selected and recomposed at runtime to satisfy the remote application logic. The platform also monitors and manages the execution of the composed microservices and updates the composition to support QoS and QoE for the end users.

3.5　Fog Node Overlay Network

FON is a dynamic overlay network established over the fog network and maintained independently by each of the fog nodes. Fog nodes in FON are connected by abstract links called fog dependency links which indicate dependency relations between microservices locate on the connected service host fog nodes (referred as FogNode for simplicity).

We define that if a microservice in FogNode α requires data or a state from Fog-Node β as a precondition for α's microservice invocation or execution, we can say α depends on β. Figure 3.5 illustrates a fog-based IoT application formed by Service A, Service B,..., Service G. The Solid line arrows show the service execution path (i.e., the dependency between subsequent microservices). The fog dependency links are represented by dash-dotted arrows and maintained in an external dependency table on each FogNode. When a microservice is newly deployed on a FogNode, the FogNode will send this microservice's precondition over the fog network to find a set of FogNodes that satisfy the precondition and establish fog dependency links to these FogNodes. In each of the FogNodes, there are service dependency links to organize microservices that co-exist in the same FogNode, represented by dotted line arrows in Figure 3.5. Such links are maintained in an internal dependency table.

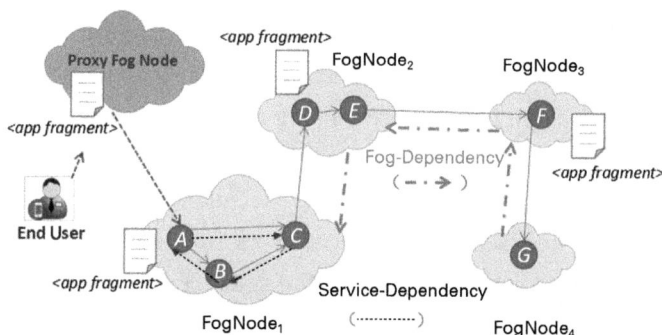

Figure 3.5　Fog-based IoT application formed by microservices

3.6　Hierarchical Microservices Management

With resource virtualization, fog services are not identified by or strictly associated with any physical asset. Instead, the services are described in a data structure and exist in a software abstraction layer named fog service overlay.

3.6.1　Fog Services and Service Composition

In this book, microservices deployed on FogNodes are modeled as a 3-tuple that includes preconditions, constraints and the effect of the service execution, which can be production of data, a controlling operation on things, or a change in the device state. Application vendors compose existing microservices dynamically to achieve a part of the software functions to reduce the deployment effort without compromising its application service coverage. To enable such microservice composition and management over fog networks, we leverage and extend a service model named GoCoMo [127] to resolve applications' functionality demands. GoCoMo is a fully decentralized service composition model that supports service discovery and execution in mobile environments. The model is goal-driven, focusing on time-efficient service provisioning to reduce the interference of topology changes. Chapter 4 will introduce GoCoMo in detail.

During execution, an application binds required microservices and then invokes them to achieve particular behaviour. GoCoMo merges the service binding with the service invocation process, which reduces interactions for service composition. It is possible that a software function could be realized by different combinations of microservices, so a service composition model is likely to have alternative execution paths for service invocation. Late service binding allows a service composition system to always select the valid execution path and prevents resources on a FogNode from being locked by an application during the whole execution process.

We extend GoCoMo's service provider network from a fully decentralized model to a hierarchical model by introducing proxy fog nodes. Proxy fog nodes are normally located close to end users and serve as caches to maintain information about local executable slices of an IoT application (a.k.a., application instances). When an executable slice of an application is deployed, the FogNode that resides somewhere easy to be accessed by end users and capable of maintaining application information can be selected as the proxy fog node for this slice.

3.6.2 Proxy Fog Nodes

A local executable slice of an application is managed and indexed by application fragments. As illustrated in Figure 3.5, each of the application fragments caches information about the composite of the subsequent microservices and indexes to the first required microservice of the composite. For example, the application fragment on FogNode1 directs to Microservice D on FogNode2, and stand for a composite microservice $\langle P_D, C_{D,E,F,G}, E_G \rangle$ including microservices from D to G, where P_D is Microservice

D's precondition, and E_G is Microservice G's effect. Note that there is no application fragment on FogNode4 because Microservice G is the last microservice in this slice. The application fragment that store the composition of all the microservice (from A to G) of this application is maintained by the proxy fog node. When invoking an application, an end user will first visit the proxy fog node to get the whole slices of the application. Proxy fog nodes for the same application are connected and cooperate to ensure seamless service provisioning even if the end user is roaming.

3.6.3   Seamless Service Invocation

In a hierarchical service composition network, admission control migrates from one FogNode to another that provides a subsequent microservice through service invocation. A FogNode controls the composition by generating a sub-goal and forwarding it to other FogNodes across the network during the resolving process, invoking a subsequent microservice through application fragments during the execution process. When a required microservice is out of reach or the end user moves to another place, the service composition network will adapt the composite of microservices to the new context.

In terms of a missing microservice, the composition process will go back to a previously executed FogNode to invoke a new execution path and replace a failed one. If there is no alternative execution path in any previous FogNode, a re-composition request is generated and resolved using GoCoMo [127]. It decouples the missed functionality into a series of sub-goals and uses backward resolution to locate the final microservice first and then the rest of the microservices. New application fragments will be generated in this backward resolving process. If the end user roams to another place and cannot access the currently invoked microservices any more before the execution result has been delivered, the end user will ask the nearest proxy fog node for this application to fetch the microservice execution result from the proxy fog network. It is possible that an end user always requests local microservices. In this case, new fragments that index those local microservices will be triggered by the proxy fog nodes, and the execution process will be migrated accordingly.

3.7   Adaptability at Edge

Given the dynamic nature of the computing environment mentioned in Section 3.2, the fog as a service architecture will not only need to be flexible enough to tackle the requirement changes, but also require adaptability at the network edge to adapt to environmental changes. Adaptation means that the edge computing system decides how to adapt without user control or with only minimal control by the user [105]. An adaptive system generally consists of two parts, including a management system and a target system. The target system is a set of configurable software/hardware resources, such as edge nodes, network bandwidth, combined microservices, etc. The management system, also known as the adaptive logic, is a set of software modules that observe the environment and manage the target system, analyze the demand for adaptation updates, plan such adaptation operations and control the execution of adaptation operations. The intelligent adaptive edge system framework usually contain a MAPE-K control framework. The MAPE-K control framework is a commonly used control method in adaptive systems [128] and consists of five main components: monitoring, analysis, planning, execution, and knowledge base [129]. The MAPE-K control framework's ideas are widely available in various types of adaptive edge computing scenarios.

The key to adaptability is to adapt the system structure and the related configurations based on the perception of the environment and the system itself. Typical perceptions are self-awareness [129] and context awareness [118]. Self-awareness is able to be aware of changes regarding system resources, and context-awareness is able to acquire information about changes that occur in the computing and physical environment while the system is executing.

In general, it is relatively easy for the system to obtain internal resource changes, which can be obtained by means of logging and setting probes. To achieve context-awareness, the system needs to use sensors to collect information about the environment and then reason based on this information [130]. When the adaptive system senses the change information, it has to analyze the relevant information, determine the severity of the anomaly, and speculate on the cause of the anomaly so that it can decide whether to take measures to compensate for it.

There are many nodes and complex functions in the target system, and it is difficult to accurately speculate the cause of the anomaly based on a small number of observed anomalies, and the emerging deep learning technology has great potential for application in this regard in recent years. Once the analysis of anomaly information is completed, the adaptive system also has to plan whether or how to compensate for the anomaly. The target systems generally involve multiple functional modules, and it is difficult to perform a complete theoretical analysis of the system. While traditional approaches rely on previous experience modeled as a knowledge base, reinforcement learning techniques have emerged in to automatically generate optimal compensation strategies.

The following sections will introduce the existing approaches that apply the whole or a part of MAPE-K control framework at the network edge to achieve adaptability.

3.7.1   Monitoring Environmental Changes

Monitoring in an adaptive system starts with setting a monitoring target, and when the monitoring target deviates from the expected state, the system analyzes the cause and finds a suitable method to compensate for it. Common monitoring targets are generally derived from the surrounding physical environment. For example, a smart home [131] system's monitoring target can be light, room temperature, noise, etc. A smart medical system's monitoring target can be the patient's heartbeat, body temperature and blood pressure [132], etc. For a smart parking system [133], the monitoring target can be the number of empty spaces in the parking lot and the distribution of empty spaces, etc.

Monitoring process can collect target-related data through sensors. If needed, new data sources can also be added to the monitoring process. Once the sensors have collected the data, they will be compared to predefined expectations to determine if there is a deviation from the expected state [134].

The operating environment of the computing task itself can also be the monitoring target. To meet the QoS and QoE requirements of computing tasks, monitoring can be divided into four tiers [135] that are system level (e.g., operating system, virtual machine), computing platform level, network connectivity level, and application level.

- System level monitoring is generally implemented through system logs and monitoring interfaces provided by relevant system drivers.
- Generally, computing platforms (e.g., Docker) come with commands to monitor their own state and utilize third-party tools (e.g., cAdvisor, Prometheus, DUCP, Scout) to provide additional monitoring capabilities [135].
- The main monitoring objectives for network connections include throughput, latency, packet loss rate, jitter, etc. [135]. Network connectivity between devices is also a common monitoring target, which can generally be achieved by sending short data packets periodically between devices.
- Application level monitoring is generally achieved by pre-setting probes in the

relevant microservices that perform specific computing tasks [134]. If the microservice developer pre-analyzes the environmental factors closely related to the task and sets up monitoring ports, the edge server can better schedule the relevant resources to meet the demand in case of resource shortage.

3.7.2 Adaptation Analysis Based on Deep Learning

The adaptation analysis process determines whether to take compensatory actions by evaluating the degree of abnormality detected in the monitoring process. In general, if the anomaly is only small random fluctuations, no action will be taken. If the level of anomalies affects the system goals (e.g., the runtime of a critical task exceeds its allowable latency limit, or system resource shortage affects the QoS), then compensatory actions must be taken. At this point, the characteristics of the anomaly are further analyzed as a way to speculate on the possible causes of the anomaly. Simple parameter anomalies can be judged by traditional numerical comparison method [119]. When there are more types of monitoring targets or monitoring targets that have many types of anomalies, more complex adaptive analysis methods are required to judge them.

In recent years, many researches have started to try to use adaptive analysis methods based on deep learning [78, 128, 136]. For example, in large warehouses, such as airports and ports, there is a need to classify and store goods according to fire safety requirements. Zhang et al. [136] proposed a way that uses Radio Frequency IDentification (RFID) technology to form the environmental information of the batch and quantity of the goods, and then uses deep learning to extract features from the environmental information and combines it with the knowledge base of previous cargo information to suggest a storage method for the cargo. The warehouse manager evaluates the suggestions and feeds the evaluation results to the knowledge base, thus dynamically updating the knowledge base.

In terms of cybersecurity, the traditional Distributed Denial of Service (DDoS) attack defense method based on statistical analysis requires manual intervention, and it cannot handle encrypted DDoS traffic. The deep learning-based DDoS method [137] solves this problem well. In addition, for zero-day exploits, traditional ECC memory-based monitoring methods require source code, which is often not possible for edge servers. With deep learning, memory can be analyzed directly and suspicious areas can be flagged [138].

3.7.3 Adaptation Planning Based on Reinforcement Learning

The goal of adaptation planning is to seek policies to compensate for detected anomalies. A common solution is to look for possible compensation policies in the knowledge base based on the causes of anomalies obtained from the adaptive analysis. If a similar anomaly has occurred before but has not been successfully compensated for, then the adaptation planning process will continue to try other possible solutions. If such a process has tried all possible solutions but has not been able to resolve the exception, then the process will decide whether to trigger an exception alert, thereby requesting other means (such as human intervention) to resolve the problem. During the operation of a pervasive application, the surrounding environment, as well as the user's goals, may change and these changes may be completely beyond what was predicted when the system was designed, so predefined compensation rules may fail. Therefore, adaptive edge systems are subject to online planning.

Reinforcement learning can theoretically generate an optimal update policy automatically and has been applied in many solutions in recent years [118, 139, 140]. Reinforcement learning is essentially trial and error learning, where the system needs to continuously interact with the environment and gradually learn the optimal mapping from the environment state to the adaptive policy in a great number of attempts.

If the environment changes, the system has to pay a certain cost before it can adjust the system's state. However, in some application domains (e.g., autonomous driving), the cost of trial and error is unaffordable. Current solutions are generally towards the specific adaptive goals, e.g., [118] proposes to use reinforcement learning techniques to design computational offloading policies based on the expected charging power and future tasks. [128] combines a knowledge base with reinforcement learning to reduce the exploration cost of new policies. [139] proposes a cooperative method that each node uses reinforcement learning to generate policies from previous experiences to gain the perspective of global information during planning process. [140] uses reinforcement learning to optimize network connectivity.

3.7.4 Strategy Execution and Knowledge Base Utilization

The policy execution process specifically executes the adaptive policy generated by the planning process. The policies include the adjustment of the system resources or parameters, such as adjusting the switch state [130] and system configuration [134], modifying the video resolution [141], and adjusting the runtime resource allocation [142].

After executing the policy, the MAPE-K control framework will continue to monitor the control objectives and continuously perform steps such as analysis, planning, and execution, and record the relevant results in the knowledge base. With the continuous operation of the MAPE-K control framework, more and more experience will be accumulated in the knowledge base, and solutions can be found faster for subsequent problems. The work of Seiger et al. [119] models the external environment and the device's own state and uses a knowledge base storing environment-aware information to pre-store the adaptive strategies corresponding to changes in the external environment or its own state. In [128], the knowledge base is pre-stored with a set of policies generated based on historical experience, thus forming an initial offline knowledge base. New policies are tried in response to changes in the environment during operation, and the policies in the knowledge base are updated. The work of Zolotukhin et al. [137] generates an initial knowledge base based on previous cargo information and sends manual confirmation feedback to the knowledge base after running adaptive analysis, thus enabling knowledge base updates.

The knowledge base can contribute to not only the planning process as mentioned but also the environmental monitoring. For example, for environmental monitoring, it is necessary to specify what level of data deviation can be considered as data anomaly. In other words, environmental monitoring needs to set appropriate data anomaly thresholds [130]. Optimal thresholds can be obtained by theoretical analysis in some scenarios, while in practice more situations require setting optimal thresholds based on experience. A knowledge base can assist the adaptive analysis process to achieve automatic tuning of this threshold in the absence of experience. To do so, a small threshold can be set at the beginning to ensure that all data anomalies are captured. Over time, more and more cases are stored in the knowledge base. The monitor gradually increases the threshold value by analyzing the previous cases to reduce the false alarm probability and reach the optimal threshold value.

3.7.5 Extension of the MAPE-K Framework

The implementation of adaptive frameworks can be divided into self-configuration, self-healing, self-protection, and self-optimization according to their specific adaptive goals [128]. Self-configuration means that the system is able to configure itself without error-prone manual installation and configuration [128]. Self-healing involves monitoring and patching tasks without human support [119]. Self-protection refers to automatic defense against malicious attacks and malicious behaviour [143]. Self-optimization is crucial if the system's performance is required to always meet the user's needs, where the application adjusts itself according to the changes in the environment [113, 16].

Depending on different adaptive requirements, modules in the MAPE-K control framework can have different focuses, and currently, new adaptive frameworks with different directions such as goal-driven and quality-driven have emerged based on the extension of MAPE-K for different requirements [144]. However, in the field of edge computing, most of the research on adaptive systems is in the area of self-configuration and self-optimization. With the development of pervasive applications and wireless communication technology, more adaptive frameworks will be needed to tackle different kinds of requirements.

3.8   Microservice Deployment and Dynamic Redeployment

The fog as a service architecture consists of three layers to meet user requirements on resource availability, architectural scalability, system interoperability, and service flexibility. These three layers extend our previous service enablement architecture [125], which are infrastructure, platform, and software layer. The infrastructure layer manages different types of devices, maintaining network connectivity among them, and visualizing the computing, networking and storage resources. Software and hardware resources on each device are abstracted as a fog node, which is made available to the platform users. The platform layer organizes the fog nodes and maintains a network of these fog nodes ("fog network") to provide a horizontal computing platform across different networks, devices, and semantic domains. It allows software vendors to deploy applications over the distributed infrastructure, which includes middleware, operating systems, and a node management module to cope with heterogeneity problems to allow software vendors to concentrate on dynamic user demands and their application's functionalities and behaviours.

The software layer is the core of the architecture. It deploys and manages microservices to meet user requirements. An application is normally a composition of multiple functionalities that satisfy user demands and can be supported by a set of microservices. For dynamic user demands, the fog nodes will cooperatively resolve the functionalities on the fly.

The software layer allows each fog node to discover the needed microservices and manage their composition and invocation. It flexibly composes microservices in a user's local computing environment to enable and guarantee seamless application execution even when the computing environment changes. In other words, when an end user moves to a new environment or when a service becomes unavailable, the software layer supports microservice re-composition to ensure application continuity.

The application deployment is a 2-step process. It includes a vendor-driven deployment with Service Level Agreements (SLA) and an on-demand redeployment. In the vendor-driven deployment step as shown in Figure 3.6(a), a software vendor uploads an application's microservices on the platform and provides a profile of potential clients to the system. The profile indicates essential non-functional service requirements like the desired QoS, the service coverage areas, and the service cost, just to name a few.

The node management module on the platform further selects a particular set of fog nodes according to the profile and deploys several copies of the microservices on the fog nodes and makes them easy to be accessed by target clients. Once the microservices are available for use, the system starts to adapt the microservice's host nodes according to the runtime service demands and the actual QoS to optimize the static deployment. In the on-demand redeployment step in Figure 3.6(b), when a user requests an application, a service request will be sent over the fog network and will be resolved cooperatively by the fog nodes. The QoS will be calculated and recorded by these fog nodes. The system uses the historical data to check if there exists a microservice that has become the performance bottleneck and decides whether or not to redeploy it. Once a redeployment is necessary, the current hosting fog node selects a new host in its vicinity, deploys the microservice and releases its own resources that were previously allocated to the microservice. As one application's microservices may not all be deployed on the same fog node, there exists heterogeneity in communications or local data types among the microservices' providers. To deal with the heterogeneity, the service invocation in the system is not restricted to a specific application workflow. We invoke microservices for applications through dynamic service composition and compose appropriate services or data converters to deal with heterogeneity. This also enables new value-added services to be provided to end users.

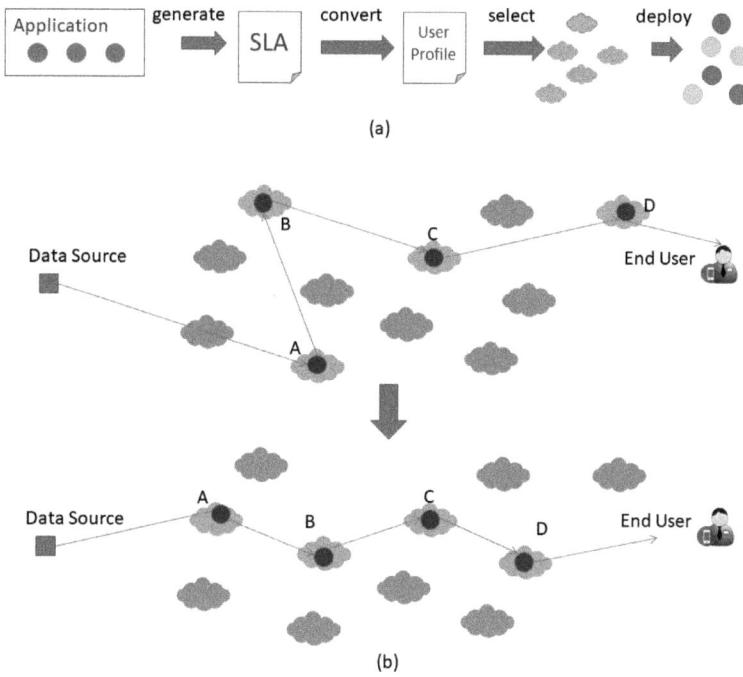

(a)

(b)

Figure 3.6 Microservices redeployment [100]
(a) Vendor-driven deployment, (b) On-demand dynamic re-deployment

3.9 Examples of Pervasive Applications at Edge

Current research has explored the value of edge computing in several application domains, such as video streaming [112], video surveillance and analysis [24], and real-time traffic route management [9]. In this section, we analyze the role of edge computing technology in practical applications through specific application examples.

3.9.1 Mobile Video

In recent years, video-based applications have gained wider attention. Watching videos on cell phones has become a daily habit for many mobile users, which brings massive amounts of data to wireless networks. According to Cisco's research report, video data accounts for 72% of mobile network data in 2019, and video-based applications are characterized by large data volume and sensitive to latency. The wireless network is susceptible to obstacle occlusion, wall reflection, multipath interference and other factors, especially in scenarios where users are in a constant state of movement and may need to switch from one base station to another, such as public transportation.

In order to improve the throughput of video transmission and reduce the transmission delay, the work of Tran [112], Liang [114], and Rahman [141] proposes edge caches to deploy mobile edge servers on base stations to cache the popular video data, thus improving the caching speed of videos for users with the same demand. A base station serves multiple users at the same time, and these users have different wireless channel quality and thus need different video quality, such as 480P, 720P, 1080P, etc. Low quality video can be obtained from high quality video by bit rate conversion, so if there is a high-quality version and a low-quality version of the video, only the high-quality version of the video needs to be cached at the edge server, thus reducing the storage footprint. However, transcoding from high quality video to low quality video consumes computing power of the edge server and may incur processing delay. The edge server needs to continuously adjust the edge cache according to the changes in the user's wireless channel quality, and [112] proposes a video resolution adjustment method that can adapt to the user's channel quality changes, which can better solve such problems. After using the edge cache, it is generally considered that the higher the hit rate of the edge cache is, the better. The work of Li et al. [144] points out that since users and edge servers are generally connected to each other through wireless networks, it also takes time to transmit the hit content from the edge cache to the users. If the transmission bandwidth between the user and the edge server is insufficient, then reducing the edge cache size will instead shorten the time required to download the video. That is, the optimal capacity of the edge cache should match the wireless transmission bandwidth.

3.9.2 Smart Home

Smart home is a typical pervasive application environment, and with the privacy issue, it's data and application logic are commonly deployed in an edge network. Unlike the traditional software control system, the smart home system is a hardware and software synergistic system. There is not only control software, but also hardware devices such as lights, sensors, switches, etc. These hardware devices may malfunction in ways that some software cannot directly sense. For example, in [119, 33], lights that can automatically adjust its brightness in order to save energy and provide a comfortable lighting system. The brightness of the lights should not be adjusted too high, but too low brightness will affect people's normal life. The smart home system can control the brightness of the lights so that the light intensity in the room is always maintained within a suitable range (e.g., 650 to 750 lux).

When the smart home system wants to turn the lights off, the smart switch will execute the "turn-off" command and reports to the smart home system that the lights have been turned off. It happens that the switch has failed and the circuit in the switch is still connected (which means the lights are still on), which is a typical software and hardware consistency problem[119]. In order to ensure that the hardware can complete the corresponding actions according to the system's instructions, after the switch reports that the lights have been turned off, the smart home system should use light sensors to monitor whether the room has become dim, and if the light intensity has not changed, the hardware component may have failed. At this point the system should attempt to fix the fault, such as trying to turn the switch off again, or turning off a higher-level switch. If the repair is successful, a record can be added to a knowledge base, and the same problem can be quickly solved when encountered again. If the fault still cannot be repaired after the attempt, a warning can be issued to request human intervention to solve the relevant fault.

3.9.3　Computational Offloading

Mobile smart devices such as cell phones and portable computers have been widely used in daily life. Due to the impact of size, power consumption, heat dissipation and other factors, mobile smart devices still have a certain gap with fixed devices such as desktop computers in terms of computing power, storage space and battery power. When dealing with computation-intensive or resource-intensive tasks, mobile devices cannot process effectively and need to upload them to the edge server, which processes and then feeds the results back to the mobile device.

In a video surveillance scenario, the camera can transmit the captured image to the edge server for data processing to reduce the battery consumption of it's own[113]. To manage the battery power, the computational tasks required for video surveillance can be further subtly decomposed into fine-grained phases such as target detection and segmentation, compression, feature extraction, and classification, each of which develops multiple algorithms ranging from simple to complex. When a computational offload is planned, a specific set of algorithm combinations is offloaded to the edge server based on the wireless channel state.

If the network connection bandwidth is limited, the camera can also perform some simple processing of the captured images before uploading them to the edge server. The test results show that the optimized computational offloading can save up to over 81% of power[113].

Many mobile devices that deployed in rural areas are powered by discontinuous energy sources such as renewable energy like solar power[38]. It is necessary for such devices to upload computational tasks to the edge server to save power when the battery is low. A more intelligent way is to compose weather forecast microservices to predict future charging conditions and reasonably plan the workload of computational tasks to overcome the problem of unstable solar and wind energy sources and extend the mobile devices' lifetime.

In addition, edge servers may also use renewable energy like solar or wind. In order to provide stable services to end devices, Xiao et al.[145] proposes a network slicing technique to unite a number of edge servers in close proximity and then dynamically allocate computing resources to each device based on available energy, user demand, QoS requirements, etc. Experimental results show that even if each edge server cooperates with only the closest server node, the total computational resources available to all edge servers for computational offloading can be doubled.

If multiple end devices simultaneously perform computational offloading to the same edge server over the wireless network, the wireless signals of these devices interfere with each other, which significantly reduces the transmission bandwidth between the devices and the edge server. If there is no interference, mobile devices can complete uploads quickly because packet loss rates are low. With network interference, the packet loss rate rises quickly and the speed of tasks uploading decreases. To reduce the packet loss rate, the mobile device has to provide sufficient signal-to-noise ratio by increasing the transmit power, which leads to a significant power consumption. The work of Chen et al.[146] investigates the multi-user computation offloading problem from a game-theoretic perspective, pointing out that a Nash equilibrium exists between these competing devices. It proposes a distributed computation offloading algorithm to achieve the Nash equilibrium.

Given the available system resources and the number of computational tasks varying over time, Lin et al.[147] proposes a distributed optimization algorithm that can adaptively adjust the computational offload policy with a long-term average response time as the goal and optimize the resource allocation according to the changing task volume. It is because many applications are not sensitive to occasional short-term performance degradation, as long as the average performance over a long period of time meets the requirements. This makes the approach impractical for task-critical or time-sensitive applications. To enable edge servers to provide acceptable services to more end devices, while to satisfy the QoS requirements of each device, Chen et al. [148] proposes a computational offloading approach that minimizes the occupation of edge servers by individual end devices.

3.10   Open Issues to Edge-enabled Pervasive Applications

3.10.1   End-to-end Security

End-to-end security is crucial for edge computing systems to provide trusted services to end users. The proposed security provisioning is anchored on certified implementations of standardized cryptographic functions and roots-of-trust. In a fog node, trusted execution environments are instantiated upon hypervisors through trusted boots to run software applications for multiple tenants. Secure communications are provided by transport-level security protocols while access control is enforced by authentication and authorization mechanisms built into data distribution services. Application and operation security including incidence monitoring and response will have to be incorporated into future extensions.

3.10.2   Distributed Runtime Management

The end devices of a pervasive application may be deployed in a scattered area of thousands of square km. The deployment locations may have completely different environments, such as indoor vs. outdoor, urban vs. suburban areas, etc. Edge servers or fog nodes are deployed close to the end devices and therefore also possibly be distributed over a large area [64]. This means that the edge systems and the pervasive applications need to be able to effectively handle dynamisms in different environments. In addition, when performing the adaptive analysis and decision making for a wide range of complex situations in a large area, it is necessary to avoid the impact of the adaptive policy of one end device on the QoS of other end devices.

A typical solution [136] is to employ an edge node with more computing resources (or even a cloud computing center) to collect monitoring data and generates an adaptation policy. However, the centralized adaptive control is difficult to avoid the negative impact of the single node failure. The execution of an adaptation policy is time-sensitive. If the communication distance between an edge server (or the end device) and the decision-making node is long, the decision results delivered to edge server will possibly fail due to further changes in the environment. Semi-centralized [116] and distributed [44, 139] adaptive control is an effective way to address the need for large scale distribution. It is still a major challenge to avoid conflicts in adaptation policies among edge services and among end devices under distributed decision making. Currently, most of the work on distributed adaptive edge computing focuses on computational offloading, and distributed adaptive edge computing in other application domains still needs further investigation.

3.10.3 Scalability and Reconfigurability

With the increasing number of end devices and edge servers deployed, pervasive applications need to have high scalability to cope with the expansion of the number of users and the increase of demand. In addition, mobile end devices can move to another place and users' demand changes over time, so the number of end devices and demands for microservices in a certain area will also change. The edge system needs to be able to adjust its structure and service provisioning plans elastically and flexibly to meet user demand and QoS requirements while avoiding idle computing resources for economic purpose. When deploying new pervasive application, the edge system needs to be able to meet its corresponding QoS requirements. This means that monitoring, adaptation analysis, planning, execution or knowledge bases associated with the new pervasive application need to be deployed to the system accordingly. As it is uneconomic to deploy these functionalities for every new application, the originally deployed adaptive system needs to be reconfigured to meet the new requirements in this case. Currently, many approaches are discussed for the increase of end devices [10, 44, 100, 131] and the growth of user volume [37, 117, 142], and less discussed for the collaboration and elastic scheduling among edge nodes [149, 150], especially lacking discussion of the existing adaptive system reconfiguration at the network edge. Therefore, systems that can satisfy multi-level elasticity requirements such as end devices, edge servers or fog nodes, and user demands need to be further explored.

3.10.4 Predictive Fault Tolerance

Predictive fault tolerance refers to the ability that the system reasoning out possible future anomalies based on the changes of the monitored target and taking measures in advance to compensate for potential anomalies. It is different from feedback-based fault-tolerant that refers to a series of remedial measures based on anomalies only after the anomalies appears.

Currently, feedback-based fault-tolerant is more common in edge-enabled pervasive applications [91, 92, 151-153]. The advantage of this way is that it is able to rapidly and accurately locate anomalies and thus perform adaptive compensation. Since the adaptive analysis process is relatively simple, it works well for adaptive requirements due to single environmental factor's changes, such as user demand changes [128], network load changes [141], and node availability changes [37]. Feedback-based adaptation's limitation is that it may cause a long waiting time and recovery process, and thus is easy to fail when used in latency-sensitive tasks.

Predictive fault tolerance can take actions in advance and is, therefore, more suitable for use in latency-sensitive tasks. Note that if artificial intelligence techniques such as deep learning are used for adaptive analysis and planning, due to its resource-rich requirement and computation efforts, reasonable algorithms need to be designed to minimize the additional latency introduced by the computation. Currently, most of the predictive fault-tolerant approaches are focused on services such as energy-saving，emission reduction [118] and computational offloading [141], and predictive adaptation for delay-sensitive tasks needs to be explored further.

3.10.5 Intelligent Edge Computing for 6G

The 6th generation wireless communication technology (6G) aims to target the needs of the next decade and will have a deep integration of technologies such as IoT, edge computing, and AI. The traditional communication network mainly completes the transmission of data, while 6G emphasizes more on intelligence. That is, the various needs of intelligent pervasive applications are considered at the beginning of design period of the communication system architecture [154]. Due to the increasing demand for computing power and the relatively high energy consumption level of AI technologies such as deep learning, it is often possible to run only lightweight AI programs directly on end devices, which is difficult to meet the needs of a wide range of intelligent applications. Supporting intelligent services in edge computing environments will play an important role in 6G.

Many AI algorithms used in industry assume that their operating environment is fixed. With AI support for wireless networks becomes stronger, more and more mobile end devices will require AI services to process their data. In mobile computing, the network topology of the system keeps changing over time, and the data collected at different locations may have differences in data distribution, making the original AI algorithm unable to correctly perform the data process task at the new location. In addition, random fading is very common with wireless channels, which can affect the communication connections between end devices and edge nodes. It will further affect the balance between synchronous and asynchronous communication in AI algorithms, and may affect data integrity. Therefore, AI applications deployed in edge systems need to be able to adapt their structure and related configurations to changes in the data to be processed.

3.11 Chapter Summary

This chapter introduces edge computing and its opportunities and open issues when supporting pervasive applications. It presents an architecture of fog computing systems to achieve a cloud-to-things continuum of microservices to enable pervasive applications. With this architecture, an end user can benefit from nearby fog nodes to receive timely services. It supports fine-grained resource management, network control, and data management.

This architecture enables fog as a service model. From an application vendor's perspective, complex user demands are decomposed into microservices, each of which does not have to maintain a global system view and can therefore run on distributed fog nodes. Cross-domain applications can also be easily developed by aggregating and composing microservices from different domains onto the same platform. This will boost end users' and application vendors' participation. From the perspectives of hardware vendors or infrastructure providers, it provides faster service responses and more efficient use of computation resources and enables new business opportunities that attract them to become edge resource providers. This architecture also enables efficient development and deployment of value-added services and supports faster iteration and response to market or environmental changes. From a network operator's perspective, it fills the computing gaps between things and the cloud, increasing the interoperability among IoT systems.

Chapter 4

Microservices Composition Model

This chapter introduces a service composition solution named **Go**al-driven service **Co**mposition in **Mo**bile and pervasive environments, GoCoMo for short [127]. It begins with discussing the design objectives and concept of GoCoMo, from which a list of requirements is presented. This chapter then presents the design of GoCoMo, and shows how the design addresses the requirements.

This chapter introduces a way to apply the design concepts proposed in Chapter 2 and builds a service composition model that supports the following required features, mapped to the challenges outlined in Chapter 2.

Feature 1: Dynamic composition planning

In a dynamic environment with no conceptual composite, a service composition model must appropriately decouple user requests to satisfy required functionality with the simplest possible composition result.

Feature 2: Self-organizing

Networked entities are likely to have only limited system knowledge, and so service providers must use local knowledge to self-organize a composition planning and execution process.

Feature 3: Minimal failure recovery delay

Composition latency is an important factor that affects the success of service provisioning and can be affected by the dynamic nature of the service provider network, which could lead to provider failures. Fault tolerance is required for dynamic environments, which handles faults that (possibly) emerge during service composition by replacing (potentially) failed service providers or a fragment of the service execution path which is invalid. A lightweight and time efficient mechanism is required to recover a microservice composite from failures.

Feature 4: Locality-driven selection

Service providers are connected through wireless communications, depending on mediators to relay messages. With more mediators for a logical service link, more message forwarding will be required, and such a link may be error-prone because of unstable wireless channels. A service composition model should reduce dependency on wireless transmission and the number of message routing hops.

Feature 5: Short standby time

Wireless links are dynamic, which may affect an established microservice composite. In particular, the longer a bound service provider has to wait for execution, the more likely a disruption to the link to invoke the service provider is to occur. GoCoMo's goal is to design a service composition model that always uses current service information to support service selection, and minimizes standby duration for selected service providers.

GoCoMo models a service composition process as AI backward-chaining. Classic AI backward-chaining processes [19], also known as goal-driven reasoning [24], have been explored to ease the resolution process in the domain of service composition. A general backward-chaining process for service composition is shown in Figure 4.1. The process starts by searching for the knowledge that can infer a request's goals (consequences), and then the request is resolved backward from the goal to the request's antecedents, by converting the goal into sub-goals, resolving back through these sub-goals (e.g., Goal: $a \rightarrow c$). The process finds a solution when all the antecedents are reached. Specifically, an initiator issues a service request $a \rightarrow d$ to start a goal-driven reasoning process, which relies on distributed knowledge bases stored in local microservice hosts (planners). In each step of the service discovery, a part of the request's goal can be solved (e.g., Goal: $a \rightarrow d$ can be partially solved on *Service*_1 that provides data transition of $c \rightarrow d$), and the remaining request (Goal: $a \rightarrow c$) is forwarded to the next-hop service providers (i.e., *Service*_2). In such a process, it is the request's goal that determines which microservices will be selected and used.

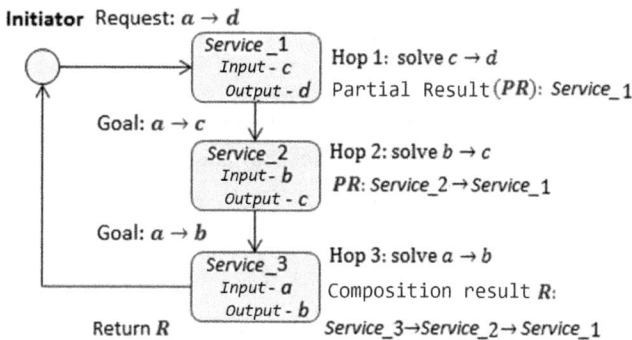

Figure 4.1   A general distributed backward-chaining model for service composition [127]

A goal-driven reasoning process produces more flexible planning results than that of workflow-driven approaches, as it considers service discovery as an open-ended problem and dynamically generates composite services according to runtime service availability.

Modelling service composition as such a process has been investigated in infrastructure-rich networks where composition planning is based on infrastructures like repositories [155] or pre-existing overlay networks [35]. However, this kind of infrastructure, as mentioned in Section 2.4, is not suitable for our target environment, and neither is the existing goal-driven reasoning processes.

To allow mobile pervasive computing environments to benefit from the flexibility that such a goal-driven service composition brings, as mentioned above, GoCoMo extends the general goal-driven reasoning model, and handles the following issues.

Dynamic goal-driven composition planning is handled via a composite participant cooperation mechanism to coordinate distributed knowledge bases and independent planners to support the generation and the maintenance of various service flows (See Section 4.2).

Heuristic service discovery is handled via a distributed heuristic discovery mechanism based on QoS attributes to increase the likelihood of time-efficient microservices being selected during execution and prevent composite requests flooding the network (See Section 4.3).

On-demand execution fragment selection is handled via online adaptable reasoning to create awareness of and compose potentially better microservices that may appear during service execution (See Section 4.4).

4.1 Microservice Model

This book assumes microservices' functions and I/O parameters are semantically annotated using globally understood semantics and languages, and able to match a service request with semantic matchmakers [122]. Such semantic service annotations are assumed to be kept in local service providers (mobile devices) and can be advertised using probe messages. A microservice's invocation must be based on all the specified input data, and local devices can form an ad hoc network, cooperating with each other to resolve a user task. Service specification is defined as follows.

Definition 1: A **Microservice** is described as $S = \langle S^f, IN, OUT, QoS^{time} \rangle$, where S^f represents the semantic description of microservice S's functionality. $IN = \{ \langle IN^S, IN^D \rangle \}$ and $OUT = \{ \langle OUT^S, OUT^D \rangle \}$ describe the microservice's input and output parameters as well as their data types, respectively. For GoCoMo, execution time QoS^{time} is the most important QoS criterion as delay in composition and execution can cause failures [23]. A service composition model should select microservices with short execution times to reduce delay in execution.

A microservice composite for user tasks can be modeled as a restrictive data transition in which the system data change from initial data (i.e., a user's input parameters) to goal data (the requested output data) while satisfying all the requested functionalities or constraints. A participating microservice for the composite, packaging its resources (e.g., data, functionality), can support all or a part of (based on its resource provision's granularity) the data transition. User tasks can be modeled as a composition request as follows.

Definition 2: A **Composition Request** is represented by $R = \langle R_{id}, I, O, F, C \rangle$, where R_{id} is a unique id for a request. The set $\mathcal{F}$ represents all the functional requirements, which consists of a set of essential while unordered functions. The composition constraints set $\mathcal{C}$ is the execution time constraints. A composition process fails if $\mathcal{C}$ expires and the client receives no result during service execution. A microservice composite request also includes a set of initial parameters *(input)* $\mathcal{I} = \{ \langle I^S, I^D \rangle \}$ and a set of goal parameters *(output)* $\mathcal{O} = \{ \langle O^S, O^D \rangle \}$.

For service providers that participate in a backward composition process, each composition request can be resolved partially (or completely), and the remaining discovery request is forwarded to their neighbouring nodes to continue the discovery process. In this composition protocol, any remaining request is enclosed in a discovery message that is forwarded between composite participants.

Definition 3: A **Service Provider** in this book is a service deployment device that has a wrapped functionality exposed through a service interface and could therefore be used remotely as a microservice.

Definition 4: A **Discovery Message** including a request's remaining part R', is represented as $DscvMsg = \langle R', cache, h \rangle$, where cache stores the progress of resolving split-join controls for parallel service flows (See **Definition 6** under Figure 4.6), and h is a criterion value for request forwarding and service allocation (For more details, see Section 4.3 and Section 4.4).

4.2   Dynamic Goal–driven Composition Planning

GoCoMo includes a goal-driven reasoning algorithm to support a fully decentralized service composition process, which is modelled as a state transition diagram in Figure 4.2 and Figure 4.5. A transition between these illustrated states is triggered by message communication events (e.g., *msgIn*, *msgOut*) or local conditions (e.g., *participate*, *usable*). Table 4.1 clarifies the notations in these figures that represent message communication events and local conditions.

Table 4.1　Composition model notations

Category	Notation	Meaning
Event	*MsgIn(x)*	Receive a message *x*
	msgOut(x)	Send out a message *x*
Message	*A = req*	Composition request
	B = dscvMsg	Discovery message
	B' = dscvMsgNew	Updated discovery message
	C = cpltTok	Complete token to represent a completely solved goal
	D = tokU	Allocation token to invoke a service
	mediateRslt	Mediate results
	E = cpltRslt	Final results
Condition	*∃ exePath*	Complete execution path exists
	allResultsIn	Requester gets all the execution results
	participate	Service hosts are ready to provide services
	usable	Services on the participant matches the request's goal
	cost	Discovery cost is still affordable (Heuristic discovery checking)
	end	An original goal has been matched
	joinService	Invoked service requires more input data to execute
	finish	The next-hop services exist
	∃ nxtService	Service execution is finished

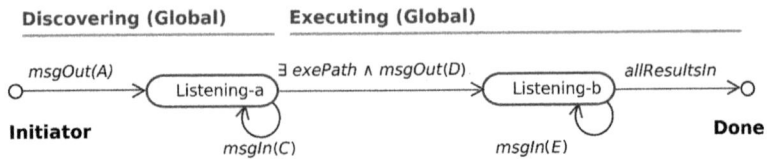

Figure 4.2　Protocol for global service composition (Client's view) [127]

Figure 4.2 illustrates a composition process from the perspective of a client (composition initiator). In particular, the global service discovery (listening-a state) starts when a client sends out a composite request to look for composite participants. Composite participants in this model are candidate service providers who are capable of reacting to and reasoning about a composite request. They are also responsible for invoking their subsequent microservices during execution. A client initializes a-Time-to-Live (TTL) value $T_{discovery}$ to manage a global service discovery process when it issues a composition request. If the $T_{discovery}$ for a composition request reaches zero, the client selects, if it is possible, a completed composite to execute.

The client then waits for execution results. Global service discovery fails if no completed composite has been received by the client when the service discovery process expires. If no execution result has been returned to the client after a previously defined period QoS^{time} before the execution process expires, the global service execution fails.

The global service discovery process establishes a temporary overlay network called dynamic composition overlay (See Figure 4.3), which contains all the reached usable composite participants. Such a network only lasts for the duration of a composition. In the network, a set of execution guideposts (See **Definition 5** below) manages composite participants and controls the discovered service flow. As shown in Figure 4.3, each guidepost is maintained by a composite participant, linking the corresponding microservice to whom sent the discovery message.

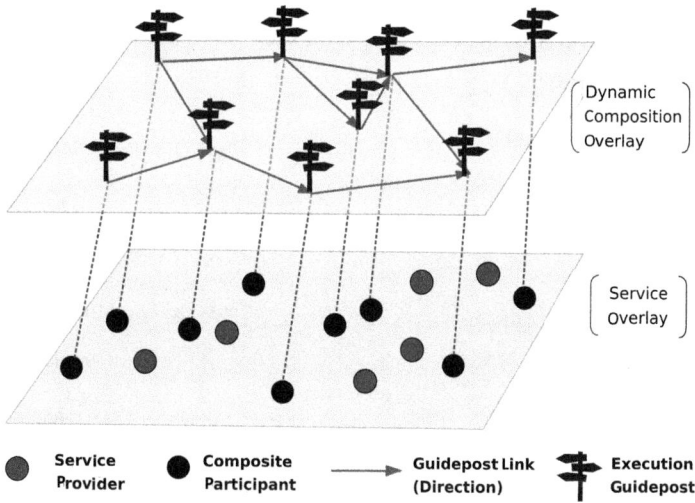

Figure 4.3　**Dynamic composition overlay** [127]

Definition 5: An **Execution Guidepost** $G = \langle R_{id},\ \mathcal{D} \rangle$ maintained by composite participant P includes a set of execution directions $\mathcal{D}$ and the id of its corresponding composite request. For each execution direction $d_j \in \mathcal{D}$, d_j is defined as $\langle d_j^{id},\ S^{post}, \omega,\ \mathcal{Q} \rangle$, where d_j^{id} is a unique id for d_j, and the set S^{post} stores P's post-condition microservices that can be chosen for next-hop execution. The set ω represents possible waypoints on the direction to indicate execution branches' join-nodes when the participant is engaged in parallel data flows. The set $\mathcal{Q}$ reflects the execution path's robustness of this direction, e.g., the estimated execution path strength and the execution time (For more details, see Section 4.4).

A brief example for GoCoMo's basic composition protocol is shown in Figure 4.4. In a network consisting of one initiator and three service providers, the initiator's composition goal can be resolved by the service providers in a decentralized way. The initiator sends a request (*req*) including the composition goal to its neighbours. The service providers resolve the goal backward by decomposing it into a list of sub-goals and matching microservices with the sub-goals. During this process, every candidate service provider creates an execution direction and stores it in its own guidepost. In each candidate service provider, such a process of resolving a sub-goal and generating an execution direction is called local discovering. After a composition goal is completely resolved, the initiator is acknowledged by the first service provider in a candidate execution path. The initiator generates an execution direction ($\rightarrow$ Provider$_1$) to keep the information about the candidate execution path (i.e., the first service provider's address, the path's robustness property, etc.). The algorithm of the local planning part is provided below.

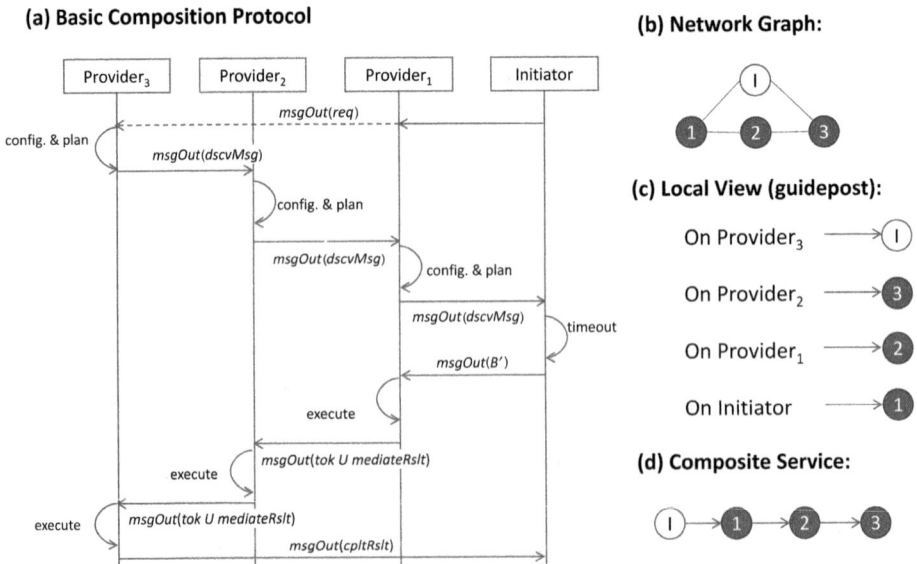

Figure 4.4　GoCoMo's basic backward planning protocol

Algorithm 4.1: Local planning algorithm

Data: Receive message *DscvMsg* from *Y*. Receiver *X* hosts $S = \langle S^f, IN, OUT, QoS^{time} \rangle$.

Result: an execution guidepost

1 /* Configuring */;

2 *GoalMatch(S, R)*;

3 /* Planning (when *usable*)*/;

4 if $\nexists DscvMsg^{\log}$ **then**

5 New *D*;

6 $DscvMsg^{\log} \leftarrow DscvMsg$;

7 if *usable* + **then**

8 *Event* $\leftarrow$ *addJoin*

9 else

10 *Event* $\leftarrow$ *add*

11 end

12 else

13 if $DscvMsg^{\log}$ and *DscvMsg* have matched or partially matched cache value **then**

14 *Event* $\leftarrow$ *addSplit*;

15 if matched **then** Remove matched *cache*;

16 ;

17 if partially matched **then** Update *cache*;

18 end

19 if $Progress(DscvMsg^{\log}) < Progress(DscvMsg)$ **then** *Event* $\leftarrow$ *adapt*;

20 ;

21 if $Progress(DscvMsg^{\log}) == Progress(DscvMsg)$ **then** *Event* $\leftarrow$ *add* ;

22 ;

23 end

24 switch *Event* **do**

25 case *addSplit*: **do foreach** $d_i \in \mathcal{D}$ **do** $S^{\text{post}} \leftarrow S^{\text{post}} + Y$;

26 ;

27 ;

28 case *adapt*: **do Clean** $\mathcal{D}$; $d_y \leftarrow \langle R_{\text{id}}, Y \rangle$;

29 ;

30 case *add* || *addJoin*: **do** $d_y \leftarrow \langle R_{\text{id}}, Y \rangle$;

31 ;

32 end

33 //When a branch's resolving is finished

34 Initiate $cpltMsg' = \langle R', cache, h \rangle$;

35 Send $cpltcvMsg'$;

When the initiator's discovery time expires, the initiator uses such information to decide which candidate execution path will be selected for execution, then the input data will be sent to the first service provider. As shown in Figure 4.4, the initiator selects $Provider_1$. $Provider_1$ gets invoked and then $Provider_2$ and $Provider_3$. The following sections will introduce the two local processes (local discovering and composing) in detail. GoCoMo names the local behaviour in service providers to react a discovery message or a request and generate the direction as local discovering, and the process on service provider that invoke and execute a microservice instance and use directions to handover execution, as local composing.

4.2.1 Local Service Planning

From a composite participant's perspective, a (local) GoCoMo composition process includes a loop for service discovery to find and link all the reachable and usable microservices, and to generate or adapt an execution guidepost. It also includes a flow for service composing process, based on the discovery result, to select, compose and execute microservices hop-by-hop in an on-demand manner. In particular, as can be seen in Figure 4.5, the local discovering loop starts in the listening state and ends when the composite participant is invoked (entering the invoking state), while a local composing flow starts in the invoking state and ends in a composition-handover state [127].

Algorithm 4.1 (Planning state) and Algorithm 4.2 (Discovery-handover state) to follow describe a local discovering loop for composite participants. This process reacts when a composite request or a discovery message is received and generates an execution guidepost as well as new discovery messages that enclose the part of the composite request which cannot be solved on this composite participant.

In the configuring state, a composite participant checks the goal matching level for a composite request (Line 2 in Algorithm 4.1). As the ultimate goal of a composite request is to produce the final, required output, if a microservice produces output that matches the one the composite requests, the microservice is usable for this composite. A microservice $S = \langle S^f, IN, OUT, QoS^{time} \rangle$ matching a composite request R's goals

$(R = \langle R_{id}, I, O, F, C \rangle\ )^{①}$ is ranked from different matching levels:

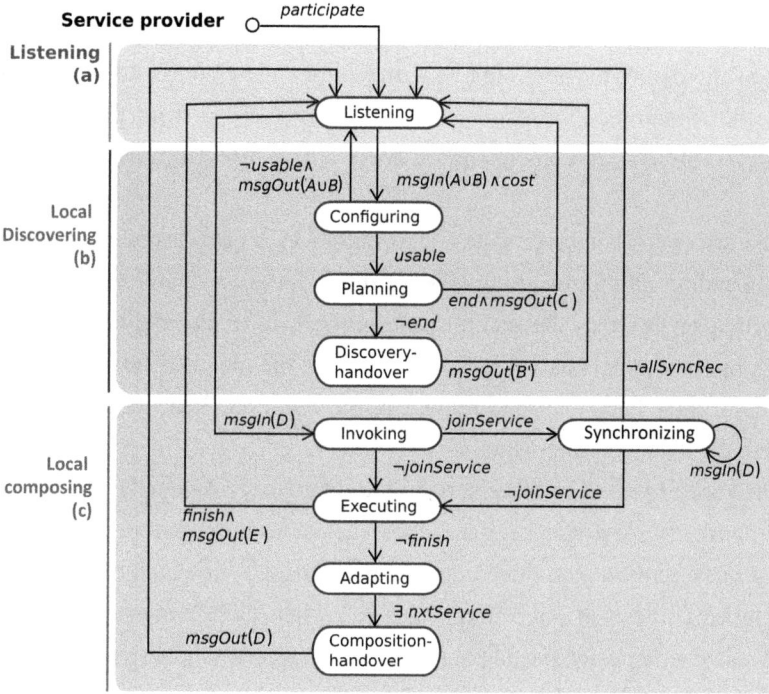

Figure 4.5   Protocol for local service composition

$$Goal\ Match(S, R) = f(x) = \begin{cases} usable & \text{if } OUT \supseteq Q \\ usable^+ & \text{if } OUT \subsetneq Q, OUT \neq \phi \\ unusable & \text{otherwise} \end{cases} \quad (4.1)$$

where *usable* represents that S supports all the goals of R; *usable+* means that S supports only a part of the goals; and *unusable* represents that S mismatches R. To prevent repeatedly checking for the same composite request, composite participants maintain a log for composite requests (Line 4 and 6 in Algorithm 4.1).

During service planning, a matched participant creates a guidepost for the composition if there is no one, and then a direction to link to the request sender (Line 5 and 24 in Algorithm 4.1) into the guidepost. Or, based on different goal-matching results, a planning process generates corresponding events to adjust the directions maintained in an execution guidepost (Line 4-20 in Algorithm 4.1).

---

① See **Definition 2** in Section 4.1 for the parameters.

For example, when a composite participant a receives a discovery message for a composition request R for the first time, and if a's microservice is usable to the composition, a will generate an execution guidepost G, model its link to the sender b of the discovery message as a direction $\langle R, b \rangle$, and add the direction into G. If another discovery message for R is received by a from the sender c afterward, and a's microservice is usable as well, a new direction $\langle R, c \rangle$ will be added to G.

In the discovery handover state (Algorithm 4.2), a new discovery message can be created depending on the matching level of the in-progress composite request. A composite participant updates the composition request, by removing the goal, adding its required input parameters as a new goal, removing the matched function and changing the execution time (QoS^{time}) requirement in the request. Then, the updated request is enclosed in the discovery message.

In the case of requiring other service providers to work together with S to support the goal, discovery messages keep a cache. The cache stores information about a node Y which sends a request R to the composite participant S, in which the goal is only partially matched (Line 8 in Algorithm 4.2), i.e., when $GoalMatch(S, R) = usable+$. In other words, a composite participant caches a request sender when the participant meets only a part of the request's goal. The cache enables composition planning based on fine-grained goals for a composition request, allowing GoCoMo to flexibly choose service providers to a composition that cannot satisfy the goal independently. This increases the scope of service discovery.

Definition 6: In a cache, a cached request sender is represented as $C_i \in cache$ $(C_i = \langle S_{id}, G^{matched}, \rho \rangle$), where S_{id} is the unique id of the requester node (a microservice, e.g., the node Y), and the set $G^{matched}$ stores matched outputs. The parameter $\rho \in (0, 1)$ captures the progress of addressing the partially matched goal (e.g., how many goals of the requests can be satisfied).

Algorithm 4.2: Local service discovering algorithm (handover)

Data: Receiver X hosts $S = \langle S^f, IN, OUT, QoS^{time} \rangle$.

Result: A $DscvMsg$ containing the remaining request R'

1 /* Handover */;

2 **if** $(Event!=add)$&&$(RemainReq)$ **then**

3 Initiate $DscvMsg' = \langle R', cache', h' \rangle$,

$R' = \langle I',O',F', C' \rangle$;

4 $O' \leftarrow IN$;

5 $F' \leftarrow F - S_{matched}^{f}$;

6 $C' \leftarrow C - QoS^{time}$;

7 Calculate h;

8 if $GoalMatch(S, R) = usable^{+}$ **then**

9 Initiate $cache_i = \langle S_{id}, m, c \rangle$;

10 $S_{id} \leftarrow P^{rec}, m \leftarrow OUT \cap \mathcal{O}$;

11 $c \leftarrow \langle \, num(m)/ num(\mathcal{O}) \, \rangle$;

12 $cache' \leftarrow cache + cache_i$;

13 end

14 Send $DscvMsg'$;

15 end

Taking a compromise goal from Anne's scenario as an example (See Figure 4.6), Anne finishes her bicycle training and is now with her 1-year-old son in a shopping mall's car park. She requires a step-free route to the nearest store that sells nappies and from where to a baby changing seat. A smart public space system in the mall includes various edge computing services owned by customers, shop clerks, taxi drivers, or housekeepers, etc. These microservices, such as GPS, navigation, translation, facility routing, taxi booking, or indoor map, can be accessed via network connections. The composite service (See Figure 4.7) allows Anne to input only a product name and a quantity number, then returns an order confirmation code to allow her to collect and pay for her order in-store. The microservice also returns an audio route stream that compacts with her smartwatch, guiding her to the store and the baby changing seat.

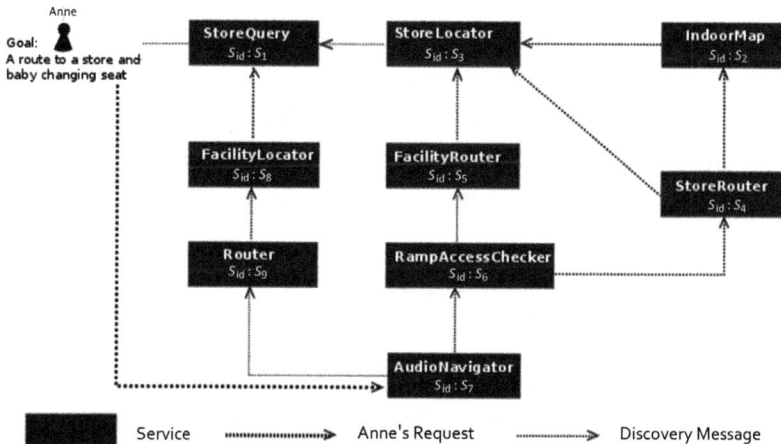

Figure 4.6   A backward planning example

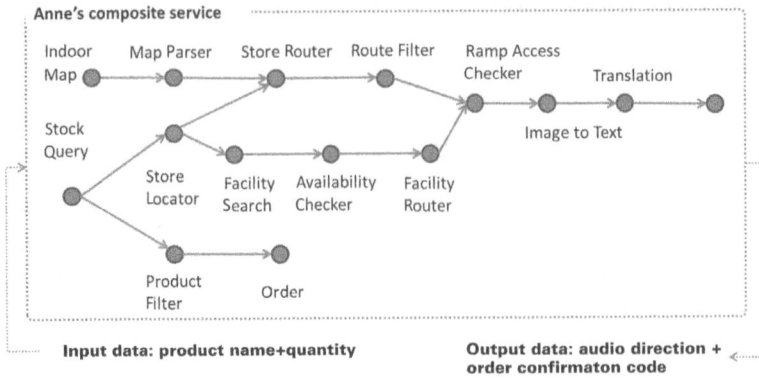

Figure 4.7   Anne's microservice composite

During the process, an Indoor Map Provider receives a discovery message from a Store Router, which includes a sub-goal for Anne to ask for two input parameters: localmap and store address. As the microservice Indoor Map can only provide the localmap, StoreRouter's goal is half matched, so IndoorMap caches the request sender by adding $C = \langle StoreRouter, localmap, 0.5 \rangle$ into the cache set. C will be forwarded, along with the discovery message, to the composite participant's neighbours until the required microservice that supports store address is found.

4.2.2   Complex Service Flows

GoCoMo resolves a data-parallel task by allowing a composite participant to recognize and generate parallel execution branches. In particular, a composite participant recognizes AND-split-join control logic for a data-parallel task and then generates a corresponding direction for it in the planning stage.

AND-split-join control logic in a composition is detected using *GoalMatch* function (4.1) and the analysis on received cache. The cache set in a discovery message also gets updated according to how the set matches the one in an old discovery message (Line 15-16 in Algorithm 4.1). A matched cache value is defined for two caches

$C_1 = \langle S_1, G_1, \rho_1 \rangle$ and $C_2 = \langle S_2, G_2, \rho_2 \rangle$:
C_1 and C_2 are

$$\begin{cases} matched & \text{if} \left(G_1 \cap G_2 = \phi\right) \wedge \left(\rho_1 + \rho_2 = 1\right) \\ partially\ matched & \text{if} \left(G_1 \cap G_2 = \phi\right) \wedge \left(\rho_1 + \rho_2 < 1\right) \\ mismatched & \text{otherwise} \end{cases} \quad (4.2)$$

As mentioned in Algorithm 4.1 (Line 7-20), in the planning state a composite participant produces an event that triggers modification on execution directions. Composition participants generate event *addSplit*, if a newly received *DscvMsg* contains a cache matching or partially matching the existing one [*matched* or *partially matched* using (4.2)]. Composite participants create event *addJoin*, if it is *usable+* to a newly received *DscvMsg* (4.1). Event *addSplit* and *addJoin* trigger the creation of two types of directions for execution guideposts:

Definition 7: An **AND-splitting Direction** directly links to multiple microservices, which requires the composite participant to simultaneously invoke these microservices for execution. An **AND-joining Direction** links to a waypoint-microservice (join-node) that collects data from the composite participant and other microservices on different branches.

Figure 4.8 illustrates how a data-parallel task is generated. In a network consisting of one initiator and three service providers, one of the service providers (Provider$_3$) depends on Provider$_1$'s and Provider$_2$'s output data for execution. During Provider$_1$'s and Provider$_2$'s local discovering process, the sub-goal generated by Provider$_3$ can neither be independently satisfied by Provider$_1$ nor by Provider$_2$. So Provider$_1$ and Provider$_2$ cache their remaining sub-goals and forward it. They also create an AND-joining

(a) Composing a parallel service flow

(b) Network graph

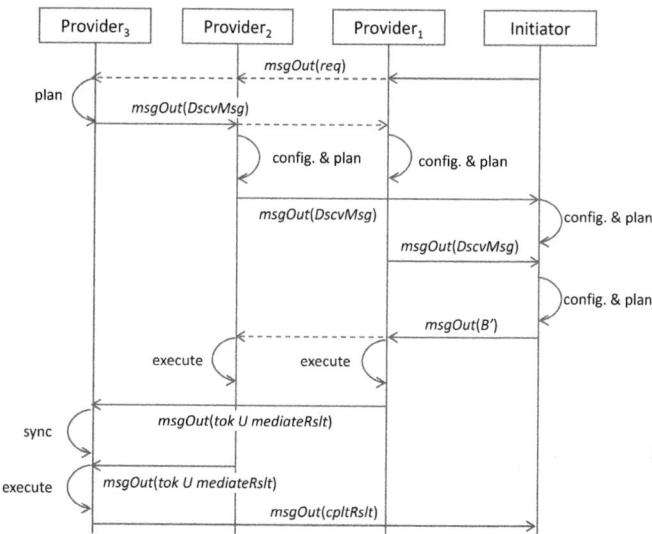

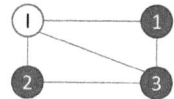

(c) Local view (guidepost)

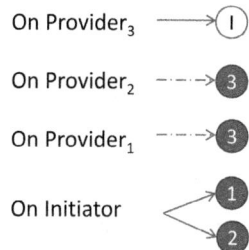

(d) Composite service

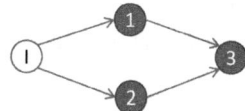

Figure 4.8   GoCoMo's backward planning protocol for composing a parallel service flow

direction that links to Provider$_3$. The initiator receives the remaining sub-goals and finds them mergeable. Then an AND-splitting direction is created on the initiator. During service execution, the execution path is split into two branches at the beginning, and then the branches are merged on Provider$_3$ by synchronizing the input data that comes from Provider$_1$ and Provider$_2$. Figure 4.9 illustrates an example in Anne's scenario for a Dynamic Composition Overlay Networks (DCON) that contains AND-splitting directions and AND-joining directions. An AND-splitting direction on the microservice **StoreLocator** links to **FacilityRouter** as well as the **StoreRouter**, while an AND-joining direction to this microservice indicates that a data synchronizing process will be needed on its subsequent execution.

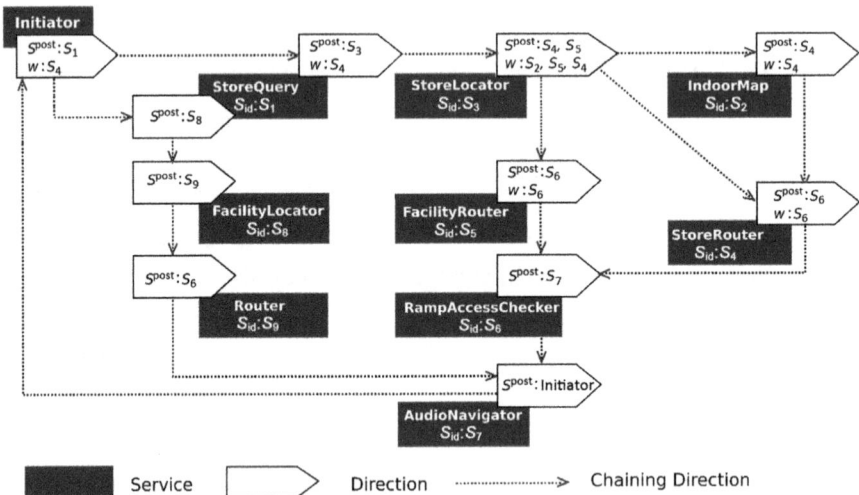

Figure 4.9   An example for DCON. S_{id} represents a unique id for a microservice

4.3   Heuristic Service Discovery

The service discovery model finds microservices hop-by-hop based on microservice data dependency. During service discovery, service providers relay a passed-in composite request when they cannot match it. Given the large number of possible data dependency relations and possible relay nodes in a network, a discovery mechanism is required. This is to find enough usable microservices in a reasonable period of time[②]

② A tolerable waiting time for a simple information query is about 2 s [156]. For operating tasks, the waiting time should be within 15 s [157].

by employing a practical number of composite participants.

GoCoMo uses a function $r(n)$ to calculate the relaying cost (relaying hops) from the last composite participant to the current composite participant S_n (i.e., the nth composite participant to resolve the composition) or a relay node R_i (the ith node to forward the composition request). These calculations are based on the heuristic value h in a discovery message.

$$h_n = \sum r(n) + n \tag{4.3}$$

The discovery cost is defined as

$$d(n) = \mu \left(h_{n-1} + r(i) \right) \tag{4.4}$$

where μ is a local communication channel parameter that is defined by local composite participants and relay nodes, which indicates the one-hop transmission delay. When a client issues a composite request, it sets up a time $T_{discovery}$ for global service discovery. When the estimated remaining discovery time $T'_{discovery}$ is smaller than a threshold τ : $T'_{discovery} = T_{discovery} - d(n) < \tau$, the node/ participant will stop relaying or processing the composition request. The threshold is determined by the remaining time constraint value on participant S_n, represented by

$$\tau = \frac{C_n}{d} \tag{4.5}$$

where d is a weight value that determines the degree of interference in service discovery, and C_n is the execution time constraint on S_n.

The interference degree d is defined according to current service density. For example, in a network that has limited available microservices for a composition, the composition model sets up a low interference degree (large threshold) to enlarge the scope of service discovery, finding more service providers for the composition. In a microservice-abundant network, a high interference degree (small threshold) is preferred to reduce system traffic for service discovery. The selection policy for setting up the interference degree and the threshold value will be discussed in Chapter 5, based on network simulation results from scenarios with different networks and service configurations. This heuristic discovery check trades off the service discovery scope with the microservice user's QoS requirements (in particular, response time). It prevents the system from over expanding the discovery scope and unnecessary communications but supports discovering sufficient composition results to satisfy the user requirement.

Consider a brief example of heuristic service discovery (See Figure 4.10). In a network including one initiator and ten service providers, the initiator wants to resolve a composition goal that has the potential to be satisfied by a composite service X. The initial service discovery time $T_{discovery}$ and the execution constraint C are 1 s. This example assumes that the time spent on one-hop-messaging is 0.1 s, and the relaying time on each participating node in a multi-hop messaging is omitted for simplicity. It also assumes that a candidate service provider needs t 0.01 - 0.05 s to resolve a goal (resolving time t) and 0.01 - 0.1 s to execute a microservice (executing time, QoS^{time}). As GoCoMo relies on backward planning, to discover X, $Provider_0$ needs to find $Service_2$ that has been deployed on $Provider_3$. According to the network graph, possible routes are: $Provider_0 \rightarrow Provider_1 \rightarrow Provider_2$ and $Provider_0 \rightarrow Provider_1 \rightarrow Provider_3 \rightarrow Provider_4 \rightarrow Provider_5 \rightarrow Provider_2$. During service discovery, every relay node checks the remaining discovery time before forwarding a discovery message, and the interference degree of the check is 1.5. As shown in the "Global View of Discovery" part of Figure 4.10, on $Provider_4$, $T_{discovery}$ equals to $1-0.1\times(1+3) = 0.6$ and is smaller than τ. $Provider_4$ stops relaying the message. This example shows how heuristic service discovery prevents a discovery message from flooding the network.

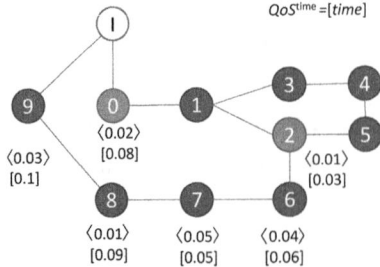

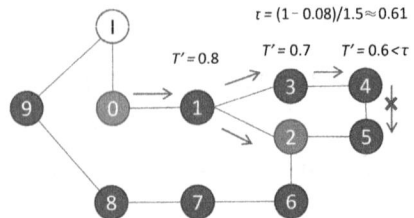

Figure 4.10   GoCoMo's heuristic service discovery protocol

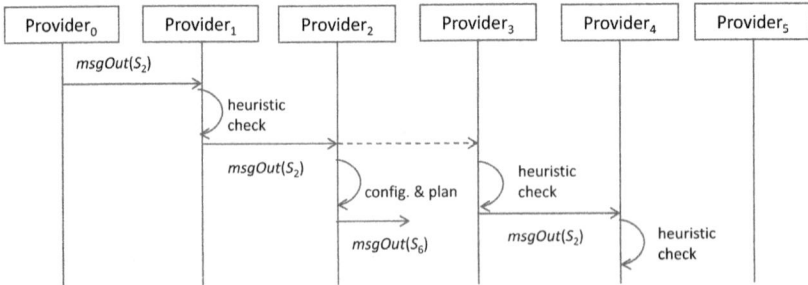

4.4   Execution Fragments Selection and Invocation

In a global view, as shown in Figure 4.11, the general goal-driven approach (a) returns a set of independent microservice composites for selection, and GoCoMo (b) uses a control element to generate a flexible microservice composite that includes runtime selectable execution paths.

Available services:

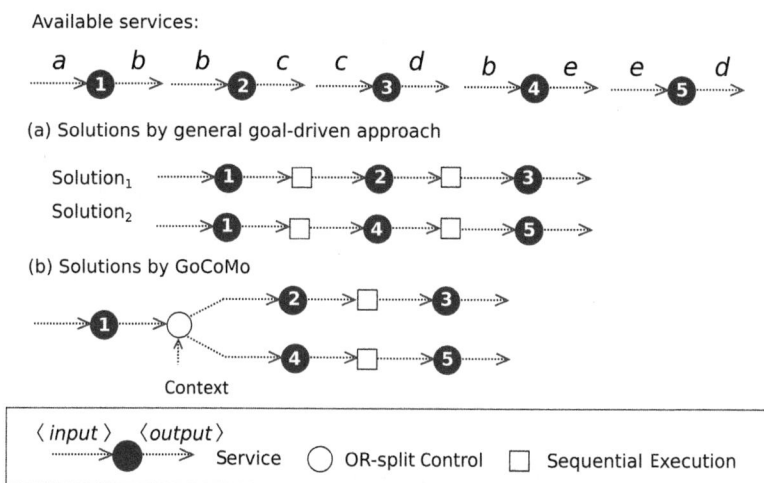

(a) Solutions by general goal-driven approach

(b) Solutions by GoCoMo

Context

⟨ input ⟩   ⟨ output ⟩   Service   ○ OR-split Control   ☐ Sequential Execution

Figure 4.11   An example of GoCoMo's execution path, in contrast to the one in a general decentralized service composition model

Control elements are enabled and managed locally by execution guideposts. Each execution guidepost that maintains possibly multiple directions acts as a OR-split[③] control element. This OR-split logic-equipped execution guidepost allows composite participants to select the best path according to the current system context, like node availability or the robustness of the remaining path. It also allows a service provider to handle a path failure by recomposing the execution path from the nearest execution guidepost other than the initiator.

A re-selection mechanism is also proposed to recover the system from path failures. Such a solution is partially found beforehand (in the global discovery stage) and can be expanded during service execution if newly matched microservices join the network.

---

③ An OR-split (split-choice) control element can have more than one outbound path, and only one of them is selected for invocation [158].

Operations on path creation and selection are managed in an execution guidepost's life cycle. Execution guideposts have a life cycle with four phases: preparing, verifying, directing, and waiting. Figure 4.12 shows the life cycle, and how it spans the duration of a participant's local discovering and local composing processes.

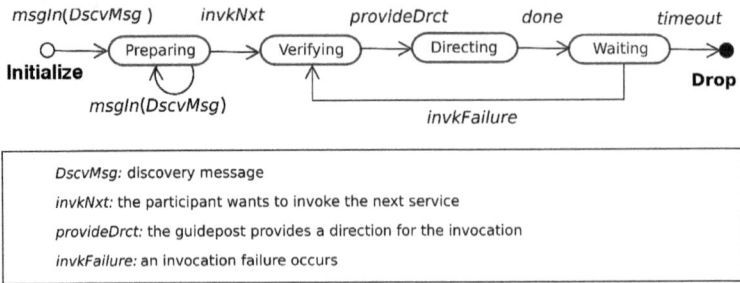

Figure 4.12   An execution guidepost's life cycle

A composite participant S_n assigns a lifetime T_n for an execution guidepost when establishing it. T_n is determined according to the remaining time constraints (C_n) and a heuristic value (See Section 4.3). Once an execution guidepost is initialized, it starts to prepare for composition by collecting directions and maintaining them. As mentioned in Algorithm 4.1 (Line 21-25), the guidepost updates and verifies the maintained directions, in responding to an event. In a local composing process, an execution guidepost fetches the address of the first service provider in a primary direction. After that the first service provider gets invoked. The execution guidepost transits to a waiting state. Subsequently, if its lifetime T_n is still not expired, the guidepost will live for a while in case the remaining execution needs a rollback for failure recovery (See Section 4.4.2). If T_n has expired, the participant drops the guidepost for the composition request, removing itself from consideration in the composition. The overall composition process continues to work with the dynamic composition overlay, as shown in Figure 4.5 (a).

4.4.1   Microservice Composite Selection and Invocation

The GoCoMo service composition process binds microservices on-demand and releases them after execution. This means a composite participant's computing resources are locked only for the duration of its local composing process. This may reduce the time a composite participant is occupied, which in turn increases its overall microservice availability. As mentioned, a client selects a microservice composite for invocation after its global service discovery process times out. Note that the actual information about the full composite is not sent to the client. Instead, the client receives the information about each microservice composite's reliability value and the first service provider's address. Based on the reliability values, the most reliable composite can be selected for invocation. The client sends a message to the first service provider to start a global service execution process. The message contains an invocation token to bind service providers and the input data for the composition. Service providers receive the message and trigger their own local service composing process, as shown in Figure 4.4 (c).

A local service composition process is a transition from the invoking state to the composition handover state (See Figure 4.5). Directions with the best quality value Q will be chosen for the next-hop invocation. Q ($Q > 0$) on participant n is calculated based on service execution time $\sum QoS_n^{time}$ on each microservice and the remaining execution path's robustness.

$$Q = \alpha^* \sum QoS_n^{time} + \beta h_n \tag{4.6}$$

where α is a weight value determined by the local network's dynamism, and h_n is a heuristic value that reflects the remaining path's length. β derives from a path duration estimation scheme [159] to estimate an execution path's robustness in mobile ad hoc networks.

$$\beta = \lambda_0 v / R_{TR} \tag{4.7}$$

where λ_0 is a proportion constant defined by network factors like node density, v is the nodes' average speed, and R_{TR} represents the transmission range [159]. GoCoMo service providers do not have a global view, and thus, such parameters are calculated using the properties of a service provider's local network (a.k.a., vicinity). Specifically, v and R_{TR} are based on a service provider's own features, and λ_0 is determined by the number of received service announcement messages (See Section 4.4.2) in a particular period. A direction that can route to a reliable execution path with a quick execution time can have a low Q value. As a direction for microservices executed in parallel may have waypoints, to synchronize a parallel service flow, the join-node will be selected when a parallel flow starts to execute.

4.4.2   Service Execution and Guidepost Adaptation

To make use of runtime proactive service announcement, GoCoMo allows composite participants to receive service announcement messages and invite newly emerged service providers to participate in a composition.

Definition 8: A **Service Announcement Message** is described as $SA = \langle P_{address},$ $OUT_p \rangle$, where $P_{address}$ represents the unique address of the service Provider, and the OUT_p is the output data that can be provided by the service Providers.

To reduce communication overhead and the time spent on dynamic service matchmaking, the service announcement message uses only the output data instead of the entire service specification defined in **Definition 1**, Section 4.1. When a composite participant receives a service announcement message from a new service provider, it uses the *GoalMatch* function (4.1) to compute if the service provider is usable to the composition. The composite participant invites usable service providers to join in the composition process by sending a previously logged discovery message. The new service provider receives the discovery message and decides whether to participate in the composition according to its local resources and service availability. If the new service provider decides to engage in the composition, it performs the local service discovery process (See Figure 4.5), adding the inviter (the composite participant who invites the new service provider) as a direction in its own maintained execution guidepost. Inviting and composing a new microservice can occur at any stage of a composition process as long as the inviter has not yet been executed.

During service execution, the availability of the first microservice on a direction is known by a composite participant through an invocation token. When a microservice is found to be unavailable, the composition model applies a back-jumping mechanism to prevent failures. During composition, a composite participant will get the id of the closest microservice that has multiple available directions from a microservice allocation token (See data D in Figure 4.5 and Table 4.1). If a composite participant cannot find a microservice available for composition handover, the composition will back-jump to the one that has multiple available directions, as long as it still locks resources for the composition (i.e., its guidepost is in the waiting phase), so that another potential remaining execution path can be picked out for execution. This process is illustrated in Figure 4.13. If the link between Provider$_2$ and Provider$_3$ is lost when executing the primary composition service Provider$_1$ → Provider$_2$ → Provider$_3$, Provider$_2$ detects such link loss and allows the service execution process to back-jump to Provider$_1$ who has a backup direction → Provider$_4$ that is able to invoke another execution path Provider$_4$ → Provider$_5$ → Provider$_3$. An execution may ultimately fail if no execution path is available.

(a) Adapt a composite service

(c) Network graph

(d) Local view (guidepost)

On Provider$_2$

On Provider$_1$

(b) Composite service

Adapt

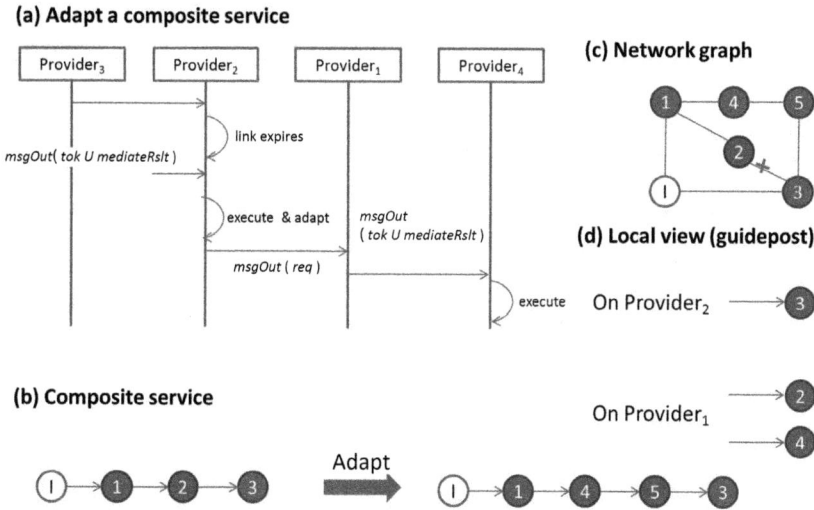

Figure 4.13   GoCoMo's back-jumping mechanism

4.5   Discussions

4.5.1   Quantitative Analysis

A quantitative analysis on GoCoMo is conducted, and the result is presented in Table 4.2. A set of parameters reflects the topics of GoCoMo's theoretical perform-ance and scalability, including the resource consumption on each participant, the complexity of composition results, how microservice density affects a service com-position process, and how many consumers or composition requests can be supported by the service composition system in a period (Workload). Generally, GoCoMo's message size and the number of messages are affected by the length of the composed service flow. A long service flow can make messages larger and requires more mes-sages to coordinate participant service providers. The composed service flow's logic and participants' mobility also have an influence on them. The more branches a ser-vice flow has, the bigger a discovery message is, and the faster the context changes, the more messages are needed to adapt the composition from any potential composi-tion failure.

Table 4.2   Quantitative analysis on GoCoMo's theoretical performance and scalability
(avg.=average)

Analysis Item	Best Case	Worst Case	Average Case
Average discovery message size	S	SL	N/A
Message/participants	2	$1 + 2N$	N/A
Overall messages	$2L$	$(1 + 2N)L$	N/A
Composition delay	$MLT \sum_{i=1}^{L} t_i$	$2MNLT \sum_{i=1}^{L} t_i$	N/A
Participant providers	L	NL	NLe^{-aD}
Workload	BN	N	N/A

N/A = Not applicable
Average size of a sub-goal request = S
Average number of candidate service-providers per sub-goal = N
Average composition length (No. of sub-goals) = L
Service distance (No. of communication hops between two neighbouring service providers) = M
Heuristic interference degree = D
Provider buffer size (the amount of composition information a service provider can maintain synchronously) = B
Time spent on one-hop transmission = T
Microservice i's execution time = t_i

Composition delay represents the time spent on executing a composite service. It depends on each microservice's execution time, the time spent on one-hop transmission, and the length of a microservice composite's execution path including the number of transmission hops between successive microservices and the number of hops to route the execution result back to the composition requester. It also relates to mobility issues, and back-jumping a composition process to recover failures caused by a missing service provider increases the composition delay. In the best case, GoCoMo can invoke a microservice composite in which any service provider is in its successive microservices' vicinity. Note that, in a real-world scenario, package loss during message transmission may increase failures and delay a composition process. The package loss rate is usually hard to be accurately and precisely predicted, and is affected by signal strength, network congestion, the receiver's message buffer size, etc.

The participating providers indicate the DCON's size, which means how many service providers join in the DCON to create a guidepost using their local resources for a service composition process. Fewer service providers for one composition can expand globally available resources for the other compositions. However, if more service providers participate in a composition, this composition has more backup execution paths when the primary one fails. The number of the participating service providers is determined by heuristic service discovery's interference degree and package loss rate. Given the use of execution guideposts and on-demand service binding, GoCoMo allows resource-rich service providers to support multiple composition requests synchronously, by maintaining a set of guideposts. The maximum workload depends on service providers' local resources (e.g., the buffer size).

GoCoMo generally reduces the failure rate of service composition in dynamic, open pervasive computing environments. Specifically, flexible compositions of services are possible by using the proposed goal-driven composition planning model, and the impact of changes in the operating environments can be reduced by GoCoMo's adaptation and execution mechanism with a reasonable cost. This section discusses open issues that have not been targeted in the previous sections but are essential in some cases.

4.5.2   Service Flows

GoCoMo is designed for mobile and pervasive computing applications that require multiple participants, and run in dynamic ad hoc environments. GoCoMo allows the generic processes of smart public spaces for pervasive computing, like multiple-source information querying, mobile sensing, data aggregation, and in-network data processing to be planned and executed in a decentralized manner. It also raises the potential for more complex functionalities like messaging-based concierge service [160, 161] to be applied in pervasive computing environments.

However, when executing a parallel service flow, GoCoMo selects a join-node before the parallel service flow gets invoked. This makes it costly to recover a parallel service flow comprising long branches when the selected join-node is missing, as re-selecting a join-node has to go back to the beginning of the branch (See the results in Section 8.3.3). In addition, as service flows that include iterative logic have not been addressed, GoCoMo does not suit applications that imply iterative transactions, such as e-commerce and intelligent control.

4.5.3   Privacy and Security

The work presented in this book relies on services that are likely to be offered by third-party providers to resolve users' request and process their data. With regard to privacy and security concerns, a trustworthy service provider network is assumed in this research, or service selection should be based on services' trustworthiness levels, represented by data from off-device authorities like reputation ratings [162] or reviews. GoCoMo does go some way towards privacy and security issues because of its backward planning model. In general, hop-by-hop processing may expose the overall control and data flow to assigned providers. GoCoMo's backward composition model partitions the data flow and the composition requester's goal. In particular, a service provider cannot see the whole data flow, and the requester's data is not visible to the full service flow's provider (i.e., the ending services' provider). In a network where individual service providers do not share any data except for those that are essential to finish a composition process, the backward composition model can prevent an entire data flow or service flow from being exposed to any service provider.

4.5.4 Semantic Matchmaking

GoCoMo discovers services with a combination of multicast routing and constrained flooding. Semantic matchmaking is assumed for service matching, though no semantic matchmaking infrastructure is required for service discovery, e.g., pre-established semantic overlay [70]. This implies that dynamic matchmaking is used, which matches local services to a received composition request at runtime. GoCoMo searches for services by dynamically matching the composition goal provided by a client to the input/output parameters of a service, but no specific matchmaker (e.g., iSeM, OWLS-SLRlite, COV4SWS, etc.) or Semantic Overlay Networks (SON) is assumed in this work. The feasibility and performance of different semantic matchmakers have been studied by [95], and are outside the scope of this book. A SON may introduce maintenance overhead to the system, however, semantic dependency overlays [122] have the potential to be used to support the discovery model described in this book, and may achieve more efficient service discovery in a particular scenario where dynamic semantic matchmaking is time intensive. However, for resource-constrained devices, the cost of maintaining a semantic overlay networks, may make participation in such an open, sharing model infeasible. A mechanism that can cope with fast and lightweight semantic matchmaking is still required.

4.5.5 High Composition Demand

GoCoMo uses planning-based composition announcements, which means service providers decide whether or not to participate in a composition process to reason about a composite and provide services. GoCoMo's service selection is based on execution path's reliability. A popular service is likely to be subscribed frequently, and so may have to maintain many execution guideposts for different compositions at the same time, such a service has, as a result, a high probability of being unavailable. If it is possible, such kinds of services should be avoided as a primary service at service binding and invocation.

On the other hand, in a network with a high composition demand, selecting a popular service for execution is probable, as a lot of service providers may be over-subscribed. It is worth exploring how to optimize the schedule of service invocation requests for different composition processes, to minimize the length of a service's unavailability.

4.6 Chapter Summary

This chapter introduces GoCoMo, and how GoCoMo tackles open issues as service provisioning failures and composition overhead. GoCoMo is a decentralized model designed for service composition in mobile and pervasive environments.

GoCoMo includes a flexible composition discovery model that supports planning-based service announcement and decentralized backward service discovery. Service providers cooperate to backward resolve a composition request from the goal to its initial state. The proposed composition discovery model generates a set of execution guideposts to enable decentralized service invocation and composite adaptation. This model also supports complex compositions. In other words, a microservice composite can include parallel service flows, when it is necessary.

In addition, GoCoMo introduces a heuristic service discovery model to achieve dynamically controlled request flooding. GoCoMo uses an infrastructure-less design for the heuristic service discovery model, and controls request floods by a threshold. Particularly, when the cost of a request routing process has exceeded the threshold, the process is stopped by service providers. Instead of assigning one uniform threshold to the global network, GoCoMo allows each individual service provider (router) to dynamically select a threshold, according to its own network property.

Moreover, GoCoMo provides adaptation and selection on fragments of execution paths (execution fragment). An execution fragment is maintained by a service provider, which includes information about the service provider's next participating provider, some important waypoints (i.e., execution branches' join node, see **Definition 7**) on this path to the final service provider, and a value to indicate the reliability of service providers on this path as well as the connection between them. Such an execution fragment is adaptable when the service provider detects a new execution path that has the potential to support the composition goal. Since execution fragments use Or-split logic to include different execution paths, when the primary path in a fragment fails, an alternative can be rapidly recomposed for execution. GoCoMo performs execution path selection dynamically on service providers. A service provider finishes its own service execution, and then selects and invokes an execution fragment for the subsequent execution.

In summary, service composition can apply the GoCoMo process to dynamically plan for microservice composites, to self-organize a composition process, to reduce failure recovery delay, and to minimize standby time for service providers, as illustrated in Figure 4.14. The remaining of this book describes how to implement GoCoMo model, evaluates how GoCoMo satisfies the design requirements outlined in this chapter, and discusses the limitation of GoCoMo.

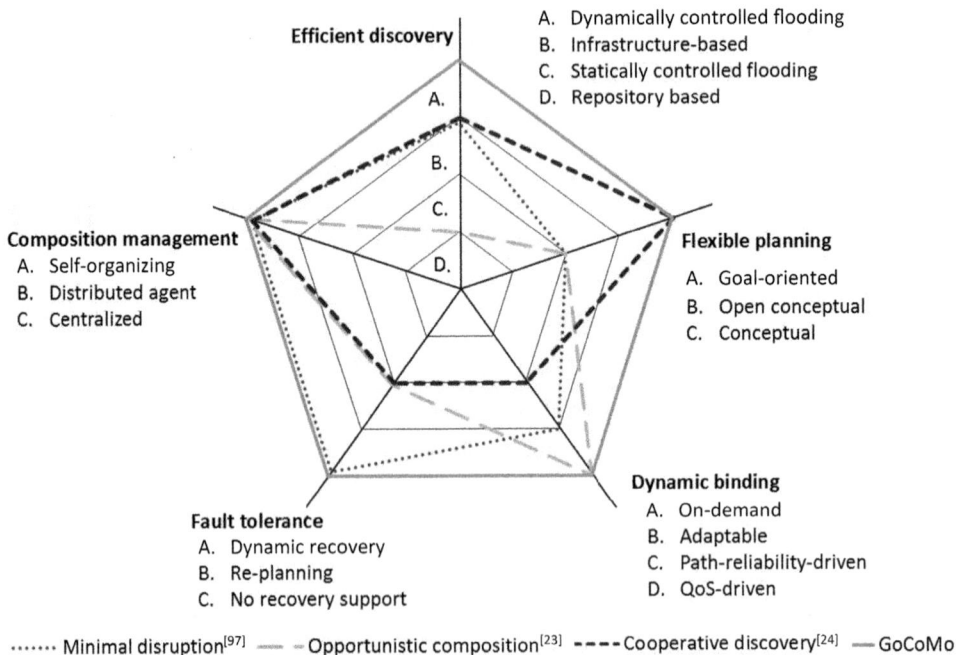

Figure 4.14   Kiviat diagram, GoCoMo's features comparing to 3 closest approaches

Chapter 5

Cooperative Microservices Provisioning

The previous chapters introduce service provisioning mechanisms on the cloud-to-things continuum. In practice, except for data collection tasks, end devices rarely play the role of service providers. This is mainly because providing services will consume their resources, but there is not yet a business model that allows them to receive equivalent benefits. However, mobile data traffic has been increasing dramatically due to the explosive growth of the mobile devices and services[163]. Such explosive data has exerted a heavy burden on current network[164]. The capacity of cellular networks should be enhanced[165], to encourage end devices to participate in appropriate service provisioning tasks.

Data-centric applications such as mobile caching is considered to be a promising solution to cope with the explosive data[166]. It caches at end devices with Device-to-Device (D2D) communication. With mobile caching, popular contents can be proactively stored in end devices' cache during off-peak time and shared directly with D2D links during the peak time, which can offload base stations' traffic and alleviate backhaul overload simultaneously[163, 164].

With a denser distribution of mobile devices[167, 168], larger cache space can be accessed nearby and a greater proportion of requests can be satisfied. By utilizing D2D communication technology, spatial reuse gain can be also achieved by constructing multiple direct links simultaneously[169]. Even with all the benefits, the number of end devices willing to participate in mobile caching is still limited, the main concern is their own limited resources and the leakage of their privacy when providing services to the public:

Limited resources

Due to the resource constraints, e.g., limited battery and cache capacity[170], end devices are unlikely to sacrifice their own resources to serve others. End devices' caching strategy tends to maximize their own profit rather than that of the overall mobile system. Such a strategy is likely to deviate from network-wide optimal policy and degrade the system performance.

Privacy

Cached contents in individual devices reflect the users' personal interests. Similar with the mobility information, it is also a kind of private information. Private information is only accessible by those authorized to view it. This makes it possible for users to circumvent contents that they are interested in, or have viewed, when providing cached data to the public. This makes the caching service provided by the local environment unable to reflect users' actual interests. Since only the content that is of interest to most local users has value for caching, caching a large amount of uninteresting data will further consume limited bandwidth resources, resulting in waste.

This chapter investigates cooperative microservices provisioning among end devices and introduces a local optimal caching algorithm that targets the limited resource and privacy issues to encourage end devices to provide caching services [171].

5.1 Cooperative Caching and Selfish Caching

Traditional mobile caching technologies, such as fully cooperative caching and fully selfish caching [172-174], cannot effectively cope with the above concerns. In a fully cooperative caching mechanism, each device selflessly shares its cached content with others and uses the complete information to formulate caching policies, which consumes a large amount of caching and D2D communication resources for content sharing and risks privacy exposure. In a completely selfish caching mechanism, each device only formulates caching policies based on its own needs to maximize the interests of itself, and the overall performance of the system will be difficult to improve when one end user's personal needs deviate from other ones.

5.1.1 Social Behaviours in Caching

Normally, devices of interest are carried and controlled by humans so that they are naturally equipped with inherent social characteristics, which open up a new avenue for mobile caching design and facilitating the solution of caching strategies. There have been lots of works introducing how to use social characteristics to save the limited resource in mobile caching [171]. In the work of Zhu et al. [175], comparing with accessing contents from others with a close relationship, devices should pay higher bids for strangers. The authors of [176] divided the local cache space of a device into its own space and the space for friends where caching strategy for the latter is just related to the interest of its friends, while the strategy of preferentially storing and forwarding the content to friends was proposed in [3]. In [177], a device can be chosen as a cache node to serve its physical and social neighbors. However, the aforementioned works consider less in protecting individuals' private information, including cached content, interest and mobility information, which are visible to all devices, including strangers'. Although privacy is an important issue in mobile caching [178], limited works focus on privacy protection in mobile caching. Jung et al. [179] proposed a privacy-preserving architecture to protect devices' location information. In [180], a privacy-preserving protocol is proposed to render such a caching system well protected against all kinds of internal or external privacy breaches. Most works like [178], [180] and [181] adopt virtual central servers to collect all devices' private information and form the caching strategy. Those virtual centralized servers do not have any social relationship and trust with devices, which also poses a risk of privacy exposure.

In order to address the challenges imposed by limited device capacities and privacy issue, the solution here adopts the concept of "social selfishness" [182] and proposes a novel mobile caching game to achieve the optimal caching strategy for mixed cooperative and selfish devices.

5.1.2　Social Selfishness of Service Providers

In human society, people tend to cooperate and show trust to their friends, but to act in an opposite way to strangers who have no or weak social relationships with them. Based on human social behaviours, we can define social selfishness as follows.

Definition 9: Social Selfishness is a mixture of cooperation and selfish behaviours of a device. An end device provides different levels of trust and service to others according to the closeness of their social relationship.

5.2　Local Optimal Caching Algorithm with Social Selfishness

This chapter introduces a Local Optimal Caching Algorithm with Social Selfishness (LOCASS) [171] for mixed cooperative and selfish devices.

LOCASS introduces a mobile caching game with social selfishness that addresses the limited resource and privacy issues in a typical scenario with mixed cooperative and selfish devices. Only cooperative neighbours trust and cooperate with each other, i.e., they share their information, capacities and contents for saving time, energy and space in local services. On the other hand, selfish neighbours do not share any resources or information for protecting their privacies.

Based on the mobile caching game, a local optimal caching algorithm is developed to achieve the best caching strategy for mixed cooperative and selfish neighbours. By introducing social-related factors, LOCASS can effectively share resources and contents among cooperative neighbours, while protecting privacies between selfish neighbours.

5.3 Cooperative Devices

LOCASS considers a mobile caching network consisting of a set of M cache-enabled devices $\mathcal{M} = \{1, 2, ..., M\}$. Those devices request contents from the library with F contents $\mathcal{F} = \{1, 2, ..., F\}$. The probability of device m to request for the content f is $p_{m, f}$. Each device m is equipped with a certain cache space with a size s_m. And each content $f \in \mathcal{F}$ has a size r_f, which can be divided into multiple segments to store in those cache-enabled devices.

Each pair of devices (m, m') is equipped with a social tie strength $\omega_{m,m'}$ which ranges in the interval $[0,1]$. Social tie strength is utilized to measure the closeness of social relationship for any pair of devices. Each device cares more about those cooperative neighbours with larger tie strengths and there is a larger probability to contribute its own cache space to store contents for them. The value of social tie strength is related to the relationship in real life. For example, the social tie strength between family members is usually larger than that of colleagues. The social tie strength of M devices are specified by the social tie strength matrix $W_{M \times M}$.

Definition 10: If a pair of devices (m, m') has $\omega_{m,m'} > 0$ and $\omega_{m',m} > 0$, they are considered as cooperative devices and will cooperate with each other in a caching task. Otherwise, they are considered as selfish devices. It is assumed that there doesn't exist $\omega_{m,m'} = 0$ and $\omega_{m',m} > 0$ for any pair of devices (m, m').

The number of cooperative devices for device m is considered as its *degree* which is represented by N_m [183]. Each pair of cooperative devices is likely to cooperate and show trust with each other while each pair of selfish devices is not. If a pair of cooperative devices is in proximity, they can be considered as cooperative neighbours, and D2D links can be established between them to share their cached contents before deadline time T_D. The system scenario is shown in Figure 5.1.

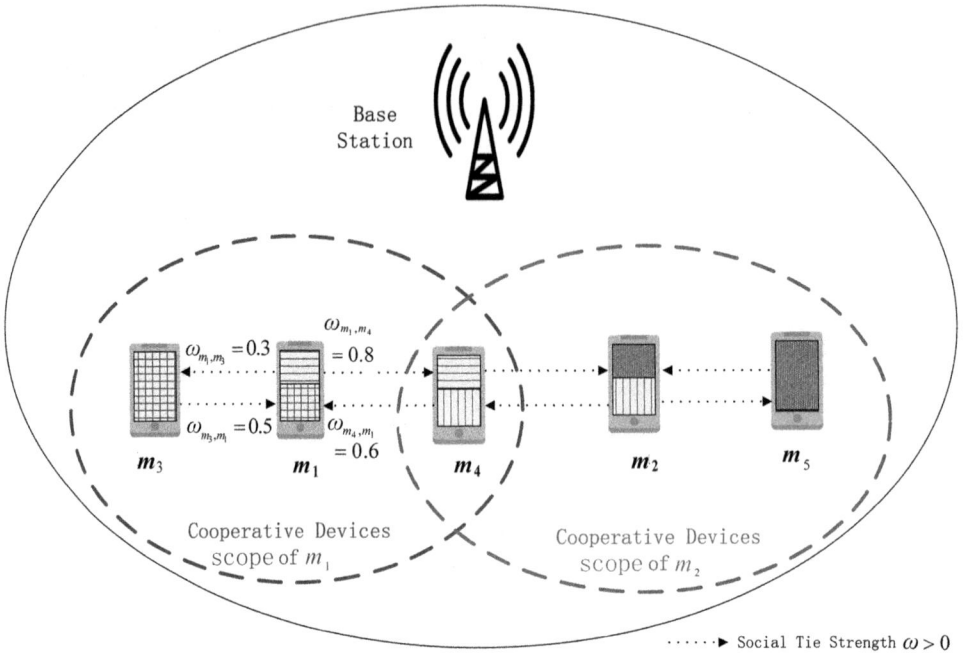

ω_{m_1,m_4}

$\omega_{m_1,m_3}=0.3$

$=0.8$

$\omega_{m_3,m_1}=0.5$

ω_{m_4,m_1}

$=0.6$

m_3

m_1

m_4

m_2

m_5

Cooperative Devices scope of m_1

Cooperative Devices scope of m_2

$\cdots\cdots\blacktriangleright$ Social Tie Strength $\omega>0$

Figure 5.1 System model, where each pair of cooperative neighbours is equipped with a social tie strength larger than 0 [171]

5.4 Social Selfishness–based Utility

An inter-contact mobility model [184] is used to capture devices' probabilistic mobility. Within a time unit, the number of contacts between the device m and the device m' obeys the Poisson process with parameter $\lambda_{m,m'}$. So, within the deadline time T_D, the expected number of contacts between the device m and the device m' is $T_D\lambda_{m,m'}$. For the device m and the device m', the expected duration is assumed to be $\theta_{m,m'}$ for a single contact. Within deadline time T_D, the expected whole D2D communication duration $T_{m,m'}$ between the device m and the device m' can be expressed as follows:

$$T_{m,m'} = T_D\lambda_{m,m'}\theta_{m,m'}. \tag{5.1}$$

With the transmission rate $r_{m,m'}$ between the device m and the device m', the device m can receive at most $T_{m,m'}r_{m,m'}$ segments from the device m' within deadline time T_D.

Content f can be recovered if the device m collects at least r_f segments within deadline time T_D. Otherwise, the device m will request the remaining segments from remote content server via the base station and the backhaul. Within T_D, for the device m, the expected amount of segments of the content f that collected from the device m' is specified by the notation $o_{m,m',f}$ which can be written as:

$$o_{m,m',f} = \min\left\{T_{m,m'} r_{m,m'} c_{m',f}\right\},\qquad(5.2)$$

where $c_{m',f}$ is the cached number of segments for the content f in the device m'. Obviously, the collected number of segments should not be greater than the number of segments cached in the device m'.

For the device m, the number of segments collected through mobile caching can be expressed as:

$$o_{m,f} = \left\{\sum_{m'\in M} o_{m,m',f}, r_f\right\}\qquad(5.3)$$

In this equation, $o_{m,f}$ should not be greater than r_f,   which means that the device m only collect the segments that just recover the content f. The concept of **offloading ratio** is defined to reflect the percent of traffic which can be offloaded from the base station via mobile caching. The offloading ratio of the device m can be represented as:

$$O_m(c) = \sum_{f\in F} P_{m,f} o_{m,f} = \sum_{f\in F} P_{m,f} \frac{1}{r_f}\min\left\{\sum_{m'\in M}\min\left\{T_{m,m'} r_{m,m'}, c_{m',f}\right\}, r_f\right\}\ (5.4)$$

The offloading ratio O_m is also considered as the **individual utility** of the device m. From the system's perspective, the average offloading ratio is defined as:

$$O(c) = \frac{1}{M}\sum_{m\in M} O_m(c).\qquad(5.5)$$

Average offloading ratio represents the percent of requests which can be offloaded from the base station and served locally via mobile caching rather than base station from the system view.

The social selfishness-based utility on mobile caching is a mixture of selfishness and cooperation caching behaviours. The social selfishness-based utility involves two major aspects, the access admission mechanism and the social group utility mechanism.

5.4.1   Access Admission Mechanism

Access admission mechanism denotes that only cooperative neighbours can be admitted to access the cached contents of a device via D2D links, while selfish neighbours cannot. With the access admission mechanism, no content sharing services are provided to selfish neighbours, which can save the limited resources. Meanwhile, devices with social selfishness are more likely to trust with their cooperative neighbours rather than selfish neighbours. Each device only allows its cooperative neighbours to access its cached contents, which can also prevent such private informa-tion of cached contents from potential exposure to selfish neighbours. For example, in Figure 5.1, device m_1 and device m_3 are cooperative neighbours with each other, and they can share cached contents through D2D links. Device m_1 and device m_2 are selfish neighbours with each other, and they cannot construct D2D links for sharing. So, the offloading ratio of the device m with access admission mechanism can be redefined as follows:

$$O_m(c) = \sum_{f \in \mathcal{F}} P_{m,f} \frac{1}{r_f} \min \left\{ \sum_{m' \in \mathcal{N}_m \cup m} \min \left\{ T_{m,m'} r_{m,m'}, C_{m',f} \right\}, r_f \right\} \qquad (5.6)$$

where the device m can only access cached contents from cooperative neighbours set $\mathcal{N}_m$. Here, N_m, the number of elements in $\mathcal{N}_m$, is also considered as the degree of the device m. For example, in Figure 5.1, N_{m_1}, the degree of device m_1 is 2. Commonly, devices with a high degree have lots of cooperative neighbours and a high social statue. The **social density** of cooperative neighbours in the system can be presented as fol-lows:

$$\eta = \frac{\sum_{m \in \mathcal{M}} N_m}{N(N-1)}. \qquad (5.7)$$

Larger η means that there are more cooperative neighbours in the system. When $\eta = 0$, there are no cooperative neighbours in the system, and each device cannot access cached contents from any other devices. When $\eta = 1$, each pair of devices can access cached contents from any other devices. For example, in Figure 5.1, the social density η is 0.4.

5.4.2 Social Group Utility Mechanism

The social group utility mechanism [185] is used to motivate devices to store con-tents for cooperative neighbours. Social group utility mechanism contains two compo-nents, i.e., individual utility and cooperative utility. The cooperative utility of device m is the weighted sum of individual utilities of the cooperative neighbours in $\mathcal{N}_m$. As mentioned before, the individual utility of device m can be regarded as the offloading ratio $O_m(c)$.

Combining the social selfishness-based access admission mechanism and the social group utility mechanism, the social selfishness-based utility of the device m can be as (5.8) and (5.9) show.

$$U_m(c) = \underbrace{O_m(c)}_{\text{individual utility}} + \underbrace{\sum_{m' \in \mathcal{N}_m} \omega_{m,m'} O_{m'}(c)}_{\text{cooperative utility}}$$

$$= \sum_{m' \in \mathcal{N}_m \cup m} \omega_{m,m'} \sum_{f \in \mathcal{F}} p_{m',f} \frac{1}{r_f} \min S(m,m')$$

(5.8)

$$S(m,m') = \sum_{m'' \in \mathcal{N}_{m'} \cup m'} \min\left\{ T_{m',m''} r_{m',m''}, c_{m'',f} \right\}, r_f$$

$$= T_{m',m} r_{m',m}, c_{m,f}, r_f - \sum_{m'' \in \mathcal{N}_{m'} \setminus m} \min\left\{ T_{m',m''} r_{m',m''}, c_{m'',f} \right\}.$$

(5.9)

For device m, taking the cooperative utility into consideration, social selfishness-based utility can motivate devices to store contents for cooperative neighbours. Considering the cooperative neighbours with different social tie strengths, device m cares more about those cooperative neighbours with a greater tie strength and there is a larger probability to contribute its own cache space to store contents for them. For example, in Figure 5.1, device m_1 cares the requests of device m_4 more than that of device m_3 due to the reason that $\omega_{1,4} > \omega_{1,3}$. Meanwhile, without taking selfish neighbours' utilities into consideration, the social group utility mechanism can also save the limited caching capacity just for cooperative neighbours' requests. To avoid exposing private information to selfish neighbours, when device m makes caching strategy decision c_m, only the mobility information $T_{m',m}$ and the request information $p_{m',f}$ for $\forall m' \in \mathcal{N}_m, f \in \mathcal{F}$ about cooperative neighbours can be accessed by the device m. Addition ally, in our proposed model, the private information about cooperative neighbour $m'' \in \mathcal{N}_{m'}$ will not be exposed to the device $m \in \mathcal{N}_{m'}$. A content requester $m' \in \mathcal{N}_m$ just need to send the quantity of required segments to the device m, i.e., $r_f - \sum_{m'' \in \mathcal{N}_{m'} \setminus m} \min\{T_{m',m''} r_{m',m''}, c_{m'',f}\}$, rather than the exact cached content information about the cooperative neighbour m'' for $\forall m'' \in \mathcal{N}_{m'} \setminus m$. For example, in Figure 5.1, the device m_4 won't leak the private information of the device m_2 to the device m_1.

Therefore, with the access admission mechanism and social group utility mechanism, the proposed social selfishness-based utility can effectively cope with the limited resource and privacy issues.

5.5 Service Deployment and Provisioning Game

With the social selfishness-based utility, a mobile caching game is proposed.

$$\Psi = \left\{ M, \{C_m\}_{m \in M}, \{U_m\}_{m \in M} \right\}. \tag{5.10}$$

The devices set M also represents the player set in the game. C_m depicts the caching strategy action space for player m that complies with the cache space size s_m. And U_m is the social selfishness-based utility of player m. In this game Ψ, each device m chooses its own caching strategy $c_m \in C_m$ aiming at maximizing its social selfishness-based utility U_m. The concept of Nash Equilibrium (NE) is introduced below.

Let $c^* \triangleq \left[c_1^*, c_2^*, \dots, c_M^* \right]$ be the solution for the game Ψ. Then the point c^* is a Nash Equilibrium for the proposed game Ψ if for any $c_m \in C_m$, the following condition is satisfied:

$$U_m(c_m^*, c_{-m}^*) \geqslant U_m(c_m, c_{-m}^*), \quad \forall m, c_m \in C_m. \tag{5.11}$$

The caching strategy of device m can be specified by the vector $c_m = [c_{m,1}, c_{m,2}, \dots, c_{m,f}]$ where $c_{m,f}$ represents the amount of the segments cached in the device m for the content f. c_{-m}^* is the caching strategy of all devices except device m. Considering the constraint on the cache size s_m, the caching strategy space of the device m can be written as:

$$C_m = \{c_m \mid \sum_{f \in F} c_{m,f} \leqslant s_m\}. \tag{5.12}$$

In this game, each device m chooses its own caching strategy c_m^* to maximize its social selfishness-based utility U_m:

$$c_m^* = \arg \max_{c_m \in C_m} U_m\left(c_m, c_{-m}^*\right). \tag{5.13}$$

5.6 Optimal Local Service Deployment

This section presents a local optimal caching algorithm with social selfishness for the mobile caching game. To design the algorithm, the existence of Nash Equilibriums should be proved. It should also be proved that each device's social selfishness-based utility is a piecewise concave function. For optimizing of the piecewise concave function, the concept of right derivative is used.

Device $m \in \mathcal{M}$ takes its own caching strategy c_m to maximize its own social selfishness-based utility U_m.

Lemma 1: for $\forall m \in \mathcal{M}$, U_m is a concave function on c.

Proposition 1:There exists pure Nash Equilibrium for the mobile caching game.

Obviously, each caching strategy c_m is a closed bounded convex set. Meanwhile, according to **Lemma 1**, for device m, the utility function U_m is a concave function on c. The utility function U_m is also continuous in c. Therefore, the caching game is a concave game [186]. By the Schauder fixed-point theorem [187], the existence of Nash equilibrium in caching game is proved.

The social selfishness-based utility of the device m can be rewritten as

$$U_m\left(c_m,c_{-m}\right)=\sum_{f\in\mathcal{F}}\sum_{m'\in\mathcal{N}_m\cup m}\{\omega_{m,m'}P_{m',f}\frac{1}{r_f}$$

$$\min\{B_{m',m,f}+T_{m'm}r_{m'm},B_{m'm,f}+c_{m,f},r_f\}\}, \tag{5.14}$$

where $B_{m',m,f}=\sum_{m''\in\mathcal{N}_{m'}\cup m'\setminus m}\min\{T_{m,m''}r_{m,m''},c_{m'',f}\}$ represents the quantity of the content f that device m' can access from the cooperative devices set $\mathcal{N}_{m'}$ or from itself except the device m before reaching the deadline T_D. When device m takes its own caching strategy c_m to maximize its utility function $U_m\left(c_m,c_{-m}\right)$, and $B_{m',m,f}$ is a constant. The utility of the content f for the device m is defined as

$$u_m^f\left(c_{m,f},c_{-m,f}\right)=$$

$$\sum_{m'\in\mathcal{N}_m\cup m}\left\{\omega_{m,m'}P_{m',f}\frac{1}{r_f}\min\{B_{m',m,f}+T_{m'm}r_{m'm},B_{m'm,f}+c_{m,f},r_f\}\right\}. \tag{5.15}$$

Thus, the right derivative of u_m^f on $c_{m,f}$ is denoted as:

$$u_{m'}^f{}^+\left(c_{m,f},c_{-m,f}\right)=\lim_{c_{m,f}^0\to c_{m,f}+}\frac{u_m^f\left(c_{m,f}^0,c_{-m,f}\right)-u_m^f\left(c_{m,f},c_{-m,f}\right)}{c_{m,f}^0-c_{m,f}}, \tag{5.16}$$

where $u_{m'}^f+$ reflects the utility improvement if an additional unit amount of the content f is stored in device m.

The function below is used to optimize the social selfishness-based utility.

$$L_{m'}^m(c_{m,f}, c_{-m,f}) = \begin{cases} 1, if\ c_{m,f} < T_{m',m} r_{m',m}\ or\ c_{m,f} < r_f - B_{m'm,f} \\ 0, otherwise \end{cases} \qquad (5.17)$$

The value of this function equals to 1 when the amount of cached segments $c_{m,f}$ can be successfully transmitted between the device m and the device m' within the deadline or help the device m' to successfully retrieve the content f. Additionally, when $L_{mm'}$ equals to 1, additional unit amount of the content f stored in the device m could bring utility improvement. We can find $L_{m'}^m(c_{m,f}, c_{-m,f})$ is non-decreasing and right continuous on $c_{m,f}$. So, given the definition of $L_{m'}^m$, $(u_m^f)_+'$ can be re-formulated in (5.18).

$$(u_m^f)' + (c_{m,f}, c_{-m,f}) = \lim_{c_{m,f}^0 \to c_{m,f}+} \omega_{m,m'} P_{m',f} \frac{1}{r_f} \frac{L_c'(m,f)}{c_{m,f}^0 - c_{m,f}}$$

$$= \sum_{m' \in N_m \cup m\ and\ L_m^m(c_{m,f}, c_{-m,f})=1} (\omega_{m,m'} P_{m',f} \frac{1}{r_f}), \qquad (5.18)$$

$$L_c'(m,f) = \sum_{m' \in N_m \cup m} L_{m'}^m(c_{m,f}^0, c_{-m,f})(B_{m'm,f} + c_{m,f}^0) - \sum_{m' \in N_m \cup m} L_{m'}^m(c_{m,f}, c_{-m,f})(B_{m'm,f} + c_{m,f}).$$

Because $L_{m'}^m(c_{m,f}, c_{-m,f})$ is a non-increasing function on $c_{m,f}$, $(u_m^f)_+'$ $(c_{m,f}, c_{-m,f})$ is a non-increasing function on $c_{m,f}$ as well. We then denote the set of the turning point $Q_{m,f}(c_{-m,f})$ below:

$$Q_{m,f}(c_{-m,f}) = \{c_{m,f} = \min\{B_{m',m,f} + T_{m'm} r_{m'm}, r_f\} - B_{m',m,f}$$

$$| \min\{B_{m',m,f} + T_{m'm} r_{m'm}, r_f\} - B_{m',m,f} > 0, \forall m' \in N_m\}. \qquad (5.19)$$

We then rank the elements in $Q_{m,f}$ in a non-decreasing order

$$Q_{m,f}(c_{-m,f}) = \{c_{m,f}^{rank_1}, c_{m,f}^{rank_2} ..., c_{m,f}^{rank_{R_{m,f}}}\}, \qquad (5.20)$$

$$u_m^f(c_{m,f}, c_{-m,f}) = \begin{cases} u_m^f(0, c_{-m,f}) + (u_m^f)' + (0, c_{-m,f}) c_{m,f}, 0 \leq c_{m,f} \leq c_{m,f}^{rank_1} \\ ... \\ u_m^f(c_{m,f}^{rank_{i-1}}, c_{-m,f}) + (u_m^f)' + (c_{m,f}^{rank_{i-1}}, c_{-m,f})(c_{m,f} - c_{m,f}^{rank_{i-1}}), \\ c_{m,f}^{rank_{i-1}} \leq c_{m,f} \leq c_{m,f}^{rank_i} \\ ... \\ u_m^f(c_{m,f}^{rank_{R_{m,f}}}, c_{-m,f}), c_{m,f}^{rank_{R_{m,f}}} \leq c_{m,f} \leq r_{m,f} \end{cases} \qquad (5.21)$$

where $B_{m,f}$ represents the number of elements in set $\mathcal{Q}_{m,f}\left(c_{-m,f}\right)$ and $0\leqslant c_{m,f}^{rank_1}\leqslant$
$c_{m,f}^{rank_2}\leqslant\ldots\leqslant c_{m,f}^{rank_{R_{m,f}}}\leqslant r_f$.

Then, $u_m^f\left(c_{m,f},c_{-m,f}\right)$ can be divided into $(R_{m,f}+1)$ parts according to $\mathcal{Q}_{m,f}\left(c_{-m,f}\right)$. So that $u_m^f\left(c_{m,f},c_{-m,f}\right)$ can be re-formulated by introducing the left deviate $\left(u_m^f\right)'_+$ in (5.21). From the equation, combined with **Lemma 1**, we can easily find that $u_m^f\left(c_{m,f},c_{-m,f}\right)$ is a continuous and non-decreasing concave piecewise linear function on $c_{m,f}$.

Based on the above theoretical analysis, a local optimal caching algorithm with social selfishness is proposed to solve the mobile caching game. The main idea of our algorithm for the caching model is to choose devices one by one and maximize the social selfishness-based utility for each of them accordingly, until the system reaches the Nash Equilibrium. In detail, for optimizing the social selfishness-based utility, the part with the largest right derivative is chosen and stored. Then, the system updates the content's right derivative to the next part and repeats the previous step until the cache is full. The detail of the algorithm is listed in Algorithm 5.1.

Algorithm 5.1: Local Optimal Caching Algorithm with Social Selfishness

1 Repeat:

2 Randomly choose one device m

3 For $\forall f \in \mathcal{F}$

4 $c_{m,f}=0$

5 compute $\mathcal{Q}_{m,f}(c_{-m,f})$

6 $count_f= 1$

7 $A_{m,f}=(u_m^f)'+(0, c_{-m,f})$

8 End for

9 $f = \arg\max\limits_{f'\in\mathcal{F}} A_{m,f'}$

10 If $\sum\limits_{f\in\mathcal{F}\backslash f} c_{m,f} + c_{m,f}^{rank_{count_f}} < s_m$

11 $c_{m,f}=c_{m,f}^{rank_{count_f}}$

12 $count_f = count_{f+1}$

13 $A_{m,f}=(u_m^f)'+(rank_{count_f}, c_{-m,f})$

14 repeat step 9

15 Else

16 $c_{m,f} = s_m - \sum_{f \in \mathcal{F} \setminus f} c_{m,f'}$

17 End If

18 Until all devices won't change their own caching strategy

19 Obtain the Nash equilibrium of caching strategy

5.7　Chapter Summary

This chapter formulates a mobile caching game with social selfishness. The social selfishness is defined as a mixture of cooperation and selfishness in this chapter. A Local Optimal Caching Algorithm with Social Selfishness (LOCASS) is introduced to prove the existence of Nash equilibrium and solve the game. Devices' mobility and request forecast are becoming more previously by adopting new technologies, such as big data. Each device transmits and receives information on mobility and request forecast of its cooperative neighbours and makes its own caching strategy decision according to the proposed caching algorithm. During off - peak time, contents are proactively distributed to the devices' cache space and satisfy the requests via D2D links at peak time.

Chapter 6

Implementation I: Service Middleware

The previous chapter describes the design of GoCoMo, and illustrates how it addresses the challenges of service provisioning in pervasive computing environments. The GoCoMo composition process is supported by a middleware, named GoCoMo middleware. The GoCoMo middleware contains the generic implementations of GoCoMo's algorithms and processes. This chapter describes in detail how to realize the GoCoMo middleware, and presents two prototypes, one for Android-based devices and the other for ns-3 platforms.

An overview of the GoCoMo middleware's architecture is presented in Section 6.1, which includes a description of the modules that comprise it, and an explanation of each module's responsibility in the GoCoMo service composition process. Section 6.2, 6.3, 6.4 and 6.5 describe the inner module implementation details and how the GoCoMo composition behaviours are realized by these modules. Section 6.6 introduces the GoCoMo middleware's prototypes.

6.1 Service Composition Architecture

The GoCoMo service composition process executes in the context of GoCoMo middleware. Figure 6.1 illustrates a high-level structure of the GoCoMo middleware, consisting of 10 modules, including 6 major modules, 3 utility modules and 1 Graphical User Interface (GUI) module. Any participant in the GoCoMo environment will install all the middleware modules. The major modules' responsibilities are listed below.

GoCoMo Client Engine (GClientE) is responsible for managing global composition processes for service clients. Its operation includes issuing composition requests, managing the discovered service composites and selecting one of them for invocation.

GoCoMo Provider Engine (GProviderE) is responsible for managing local service composition processes for service providers, through interacting with other modules to manipulate such a service composition process's state transition.

Execution Guidepost Manager (ExeGM) underpins **GProviderE** by managing the information about the discovered service providers and service composites. It is responsible for overseeing the local execution guideposts' life cycles and working out the best execution path for service invocation.

Control Logic Helper (CtrlLogic) is responsible for deciding if a composition request matches a local microservice and how the local microservice can engage in the composition. **CtrlLogic**, if necessary, also reasons about control logic in a service execution path, such as AND-split-join. In other words, it permits **ExeGM** to include parallel microservice flows to an execution path.

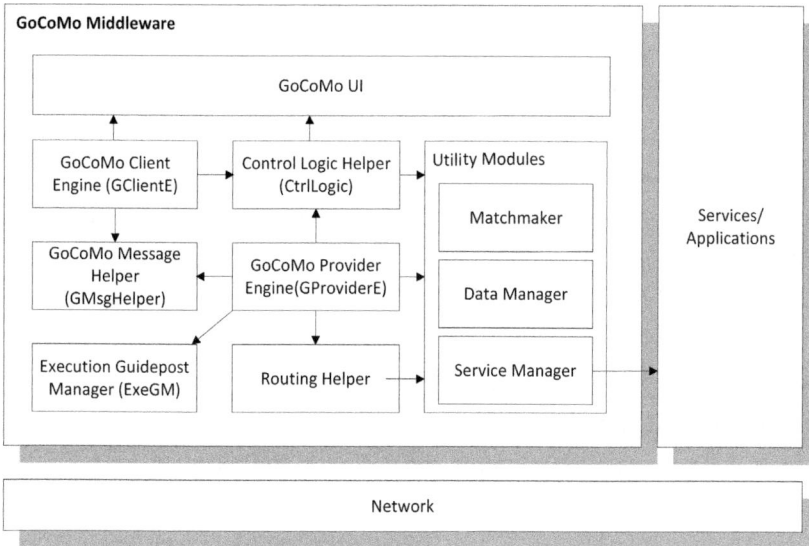

Figure 6.1   The GoCoMo middleware's architecture: 1 UI module, 3 utility modules and 6 major modules. (Arrow lines represent interactions between these modules)

Routing Helper assists **GProviderE** to determine whether to continue a request routing operation, which implements the heuristics service discovery model introduced in GoCoMo.

GoCoMo Message Helper (GMsgHelper) is responsible for parsing or generating messages for information exchange between composite participants. Such messages include composition requests, *DscvMsg* (discovery messages), complete tokens, invocation tokens, recovery token, and service announcement messages. A set of utility modules and a GUI module in the GoCoMo middleware collaborate with the major modules, facilitating composition processes and user interaction.

GoCoMo UI is a user interface that obtains user requirements from human users for **GClientE** to issue a composition request, and receives microservice registration information from microservice providers.

Service Manager provides an interface for microservice-related operations. Specifically, it allows microservices to be registered with the GoCoMo middleware on its own device, to enable service provisioning. It also invokes local microservices for execution at runtime, and fetches execution outcomes from the invoked microservice instance.

Data Manager is responsible for caching/fetching support data for a composition process, such as historical discovery messages and network property data.

Matchmaker is responsible for matching two entities and returning the match-making result to **CtrlLogic**. Matchmaking method can be syntactically or semantically depending on the domain of the operating environment.

These GoCoMo middleware modules collaborate with each other to search for service composites on behalf of composition clients, and let mobile devices participate in a composition process as a service provider.

From a composition client's perspective (i.e., composition requester), the Go-CoMo middleware receives a composition requirement from **GoCoMo UI** and formats the requirement as a service composition request, using **GMsgHelper**. The request is then sent to the local network to initialize global service discovery process. Before the service discovery process finishes, any discovered service composite pushed to the client will be verified by **GClientE**. **GClientE** keeps only the valid ones. After the discovery process, **GClientE** selects a service composite from those stored previously, and invokes it by sending the required input data to the first service provider in the composite, and then waits until an execution result arrives.

The GoCoMo middleware shares a provider's services in two ways: planning and advertising. In a planning process (See Section 4.2), a composition request, parsed by **GMsgHelper**, is firstly checked by **CtrlLogic** to find out whether a provider's local microservice satisfies the composition request. In particular, **CtrlLogic** uses Match-maker to match the composition request's goal to the local microservice's output, and returns matchmaking outcomes to **CtrlLogic**. If the outcomes show that the microser-vice is "usable", **CtrlLogic** generates an event that suggests how the provider can engage the composition. **GProviderE** gets the event from **CtrlLogic** and asks **ExeGM** to initialize or update an execution guidepost, depending on the event.

A provider device advertises a microservice using **Service Manager**. If a service provider that engages in a composition process as a participant, receives a service ad-vertisement message, the participant's **GMsgHelper** draws service information from the message. Then, **GProviderE** fetches a locally cached passed-out discovery mes-sage from **Data Manager**, and calls **CtrlLogic**, examining if the microservice matches the composition's goal. As mentioned in Section 3.4, the matched microservices' pro-vider gets invited to join in the composition by the participant.

During service execution (See Section 4.4), **Service Manager** invokes a local mi-croservice, and passes the input data to **GProviderE**. After the local microservices are executed, a direction that indicates the subsequent execution path is selected by **ExeGM**, and **GProviderE** sends the execution result to the next service provider in the execution path. The service execution process will be continuously performed by the subsequent service providers.

More details about the interaction between these GoCoMo middleware modules and the inner module implementations will be presented in the rest of this chapter.

6.2   Client and Provider

GoCoMo App, as mentioned, has **GClientE** and **GProviderE** modules that realize the management of global service composition and local service composition, respectively. This section describes the inner implementation of these two modules, and their interactions with other GoCoMo middleware modules.

6.2.1   Client Engine

GClientE is a group of classes that enable GoCoMo's global service discovery and execution for composition clients (requesters). A global service discovery and execution process (See Figure 4.2) includes the behaviours of issuing composition requests, receiving composition results, selecting a service composite from the results, and invoking the selected composite for execution. **GClientE** deals with all aspects of the client behaviours in a GoCoMo composition process.

Figure 6.2 illustrates the class diagram of **GClientE**. **GClientE** implements two classes: **GoCoMoClient** and **CompositePool**. A **GoCoMoClient** object receives a GoCoMo message and conducts overall global composition management. It includes an active() method to atomically execute service searching or service composite updating. **CompositePool** extends the **Guidepost** class to get access to and maintain the composite data. That data is expressed as a **Direction** object, the implementation of which will be detailed in Section 6.4.

Figure 6.3 illustrates the major interactions between **GoCoMoClient**, **CompositePool**, and the classes involved in a global service discovery and an execution (invocation) process. When the GoCoMo middleware receives a GoCoMo composition request, a **GoCoMoClient** object is created, and at the same time a TTL value (See Section 4.2) is assigned to the object that represents the object's lifetime. **GoCoMoClient** initializes a **CompositePool** instance, and calls a *send* () method in **SendService** class to pass the request to other peers in the network.

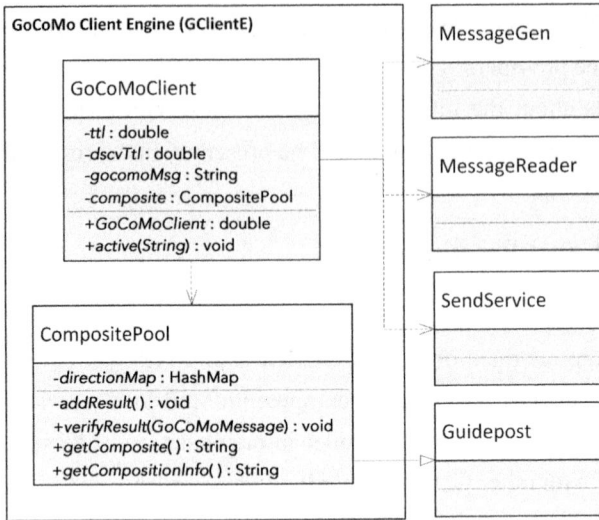

Figure 6.2   GoCoMo client engine implementation: class diagram of GoCoMo client engine and its relations to other GoCoMo modules

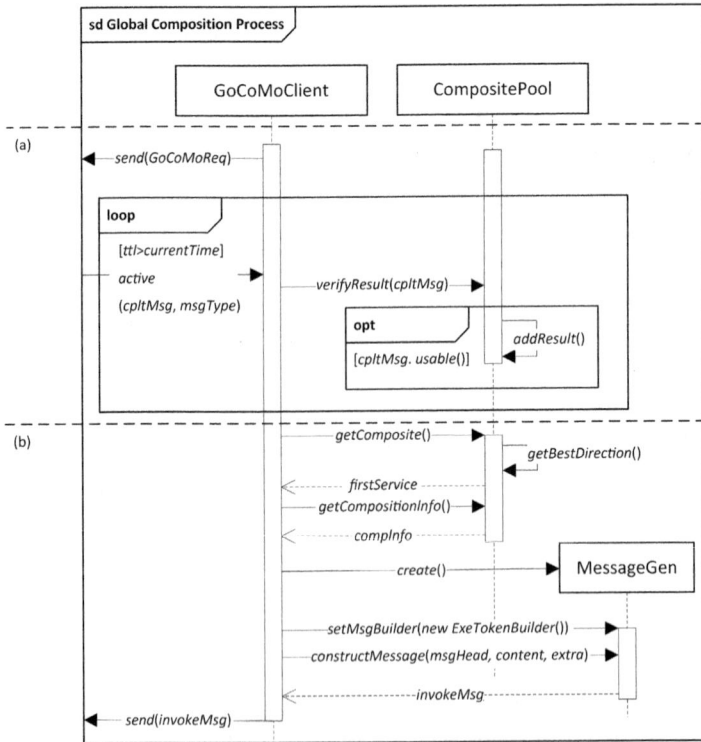

**Figure 6.3   Global service composition process implementation
(a) High level sequence of global service discovery, (b) Invocation**

GoCoMoClient receives cpltMsg from a service provider that contains information about a discovered service composite, if the request is resolved by the service provider. **GoCoMoClient** retrieves a composite data, and calls a *verifyResult()* method in **CompositePool** to check the validity of the composite (i.e., whether or not this composite satisfies all the functional and non-functional requirements in a GoCoMo request). **CompositePool** keeps valid composites by modelling them as **Direction** data and storing this data in a key-value pair directionMap.

After the global discovery process times out, **GoCoMoClient** invokes the *getComposite()* method in **CompositePool** to obtain the address of the first service provider in the best candidate service composite. It then fetches extra information about the composite, for example a waypoint service provider in the execution, using the *getCompositionInfo()* method. **GoCoMoClient** calls a message object in **GoCoMoMessage** class to obtain a *invokeMsg*, getting ready for the composite's execution.

6.2.2   Microservices Provider

The GoCoMo Provider Engine (**GProviderE**) is implemented as a **GoCoMoProvider** class. Instances of **GoCoMoProvider** locally control all the request resolving and microservice routing behaviours. The GoCoMo middleware invokes a **GoCoMoProvider** instance to handle a composition request, and launch a local service composition process. Figure 6.4 shows how a **GoCoMoProvider** object interacts with other GoCoMo classes to enable local service composition processes.

In local service discovery, a composition discovery message (*DscvMsg*) received by the GoCoMo middleware is first managed by a **GoCoMoMessage** object that analyzes its type, and retrieves the content of this message. The processed message is then passed to **GoCoMoProvider**, as Figure 6.4 (a) shows. **GoCoMoProvider** first calls *genLogic()* in **LogicController** class that returns an event data to trigger the locally cached execution guidepost's adaptation. As mentioned in Section 4.2, GoCoMo defines four events, each of which triggers an adaptation behaviour defined in **GuidepostManager**. Adaptation behaviours include *add, addJoin, addSplit* and *adapt*. The event data is sent to an instance of **GuidepostManager** to generate/update an execution path. **GoCoMoProvider** afterwards chooses to keep forwarding the remaining request or to send a discovered composite to the client, depending on whether or not the passed-in request has been completely resolved.

Service advertisement messages are also handled in local service discovery processes.

As shown in Figure 6.4 (b), **GoCoMoProvider** receives a service advertisement from other devices, asks **LogicController** to measure if the microservice's output meets the composition's goal, and then invites the provider of the microservice that satisfies the composition's goal to engage in the composition by sending out a *DscvMsg*.

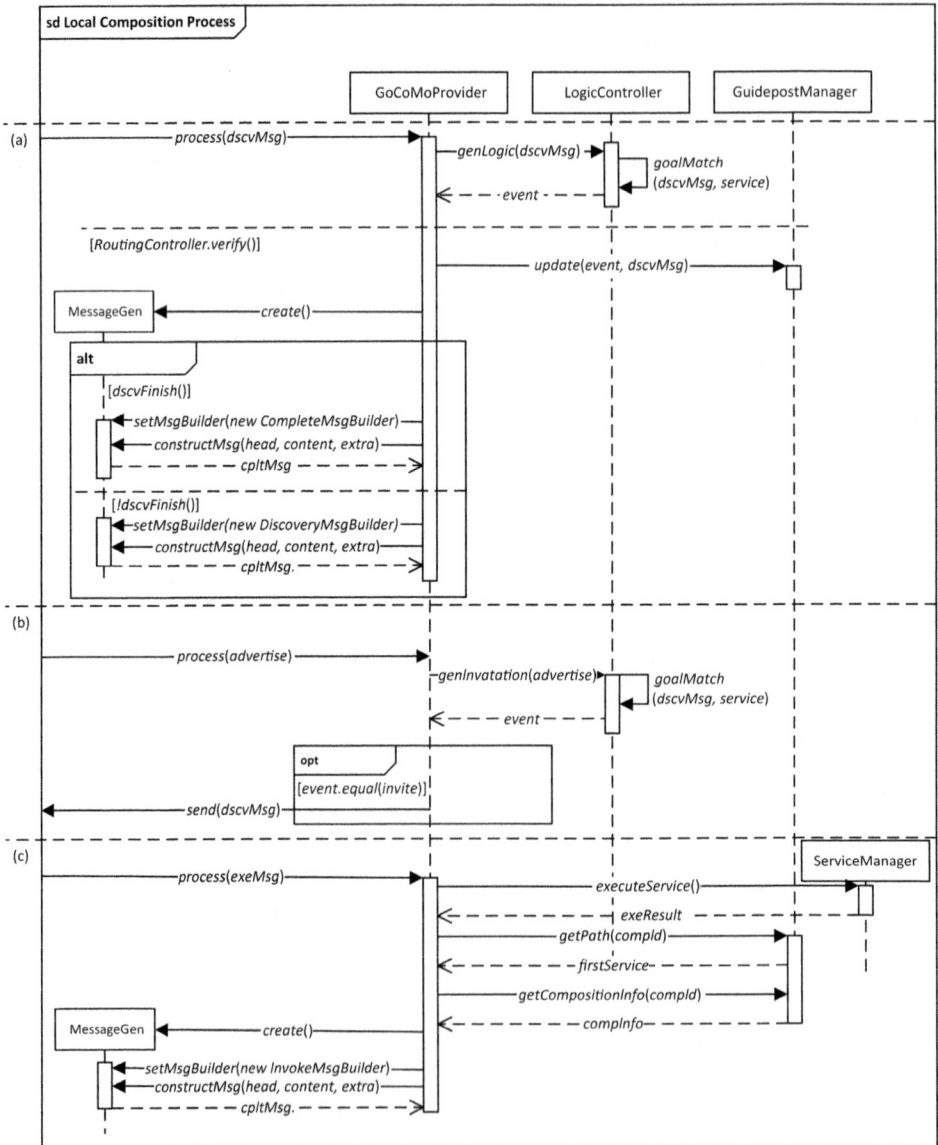

Figure 6.4   Local service composition implementation
(a) High-level sequence of local service discovery, (b) Inviting new service Providers, (c) Invocation

When a global service execution process reaches a local microservice, this microservice's execution is invoked by an *exeMsg* data. As introduced in Section 4.4, the *exeMsg* data includes the input data of the local microservice and the address of the last guidepost that maintains multiple directions for this composition. Besides, if such a microservice is in a parallel execution flow, the address of its join node (waypoint) is also included. Figure 6.4 (c) shows how **GoCoMoProvider** invokes a local microservice and how the subsequent service composite is selected. Similar to **GoCoMoClient**'s behaviour shown in Figure 6.3, **GoCoMoProvider** obtains the next service provider's information and receives an *exeMsg* in the get*ExeMsg*() callback. The *exeMsg* is then sent out to invoke the next service provider.

6.3　Routing Controller

In a local service discovery process shown in Figure 6.4(a), **GoCoMoProvider** uses a *RoutingController.verify*() method to get permission to continue the local service discovery process. The **RoutingController** class, as shown in Figure 6.5, has methods to compute discovery cost and generate new heuristic values. The *verify*() method in **RoutingController** returns a Boolean value, and a "true" value represents that the discovery cost is still affordable and the local service discovery may continue. Heuristic values were introduced in Section 4.3. See (4.3) for more details.

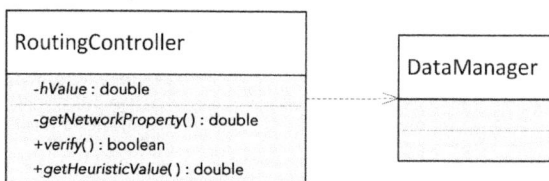

Figure 6.5　Class diagram of RoutingController

6.4　Guidepost Manager

A guidepost is a networked element in a dynamic composition overlay network (See **Definition 5** in Section 4.2). A guidepost is associated with one GoCoMo composition process, caching information about candidate service composites. Such candidate service composites' information is modeled as directions. Each direction has a unique id, and contains a list of service providers that would require input data from the guidepost's host device for service execution, a list of way-point nodes, and a value that indicates the reliability of its corresponding service composite (See **Definition 5** and **Definition 7** in Section 4.2 for more details).

The GoCoMo middleware uses Guidepost Manager to oversee a guidepost's life cycle. Guidepost Manager is a group of classes supporting a series of behaviours including creating execution guideposts, verifying a guidepost, updating directions, etc.

Figure 6.6 depicts Guidepost Manager's class diagram, which contains classes **Guidepost**, **Direction** and **GuidepostManager**. **Direction** is a complex type containing id, postConditionNode, QoS, etc. **Direction** objects for the same composition are maintained in a List, distinguished by their id attributes, and mapped to one **Guidepost** object. In other words, **Guidepost** objects and **Direction** objects are saved as paired key value sets. Once a **Guidepost** object is destroyed, all the **Direction** objects maintained in the **Guidepost** will be dropped. **GuidepostManager** consists of methods that get data from or update a **Guidepost** object.

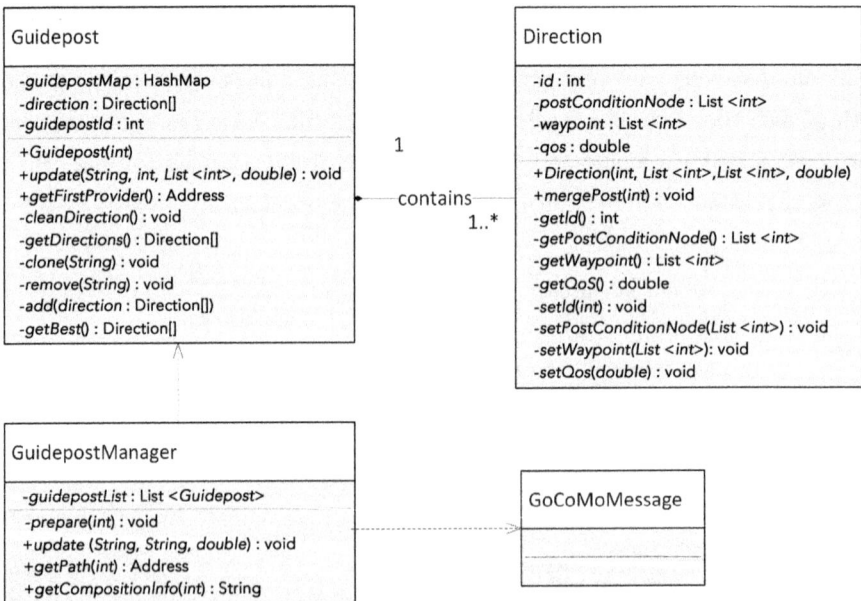

Figure 6.6   GuidepostManager implementation: a class diagram of GuidepostManager

6.4.1   Adapting a Guidepost

A key point for GoCoMo is that a guidepost can be dynamically adapted, which allows a participant to cache newly available service providers for a composition, to locally merge multiple execution branches to form a parallel service composite, and to replace service providers by one that provides better QoS. As mentioned in Section 6.2.2, GoCoMo Provider passes an event (*add, addJoin, addSplit* or *adapt*) to **Guide-postManager** to trigger adaptation on an execution guidepost as shown in update (event, *DscvMsg*) in Figure 6.4(a). Figure 6.7 outlines diverse events and their corresponding adaptation behaviours.

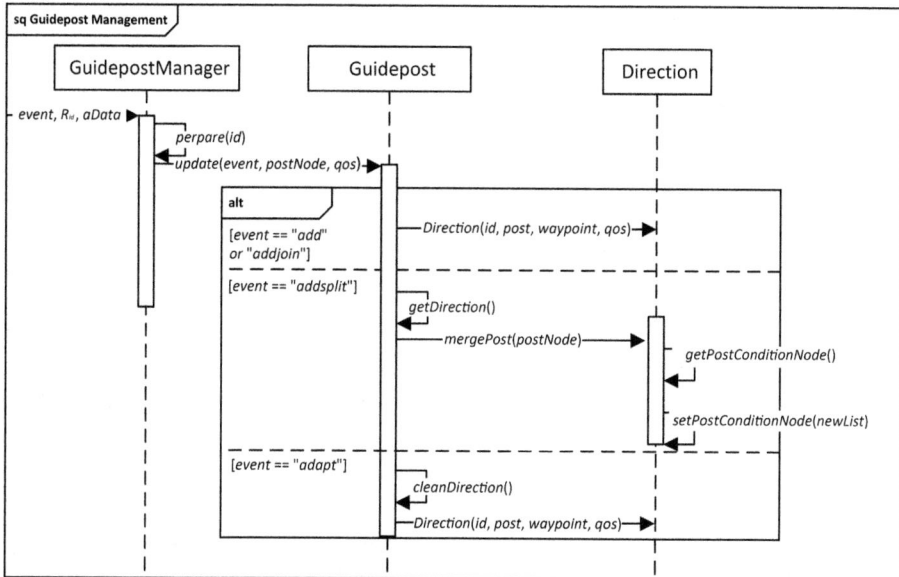

Figure 6.7　GuidepostManager sequence: high-level sequence of Guidepost adaptation behaviours. (Interactions related to the HashMap object are not illustrated)

GuidepostManager receives this event, a composition process id and the adaptation data, and then fetches the corresponding **Guidepost** using the composition process id. The **Guidepost** gets activated, and updates the direction it maintains based on the event. For event *add* and *addJoin*, **Guidepost** obtains the **postCondition**, the waypoint list, and the QoS from the adaptation data, and uses them to create a new **Direction** object and a key-value pair to store the object. Note that **Guidepost** inserts the post condition node into the waypoint before generating direction (See **Definition 7**, AND-joining direction) for an *addJoin* event. For event *addSplit*, **Guidepost** updates currently stored directions by adding a new node into their *postNode* list. For event *adapt*, **Guidepost** calls the *cleanDirection()* to remove the stored **Direction** data, and then creates a new **Direction** mapping to the composition's **Guidepost**.

6.4.2 Guidepost Data in Service Execution

A **Guidepost** object is used in a service execution process to provide composite information for **GoCoMoProvider**. The main operations to get composition information from **GuidepostManager** have been introduced in Section 6.2.2, and depicted in Figure 6.4 (b). In the **GuidepostManager** class, the *getFirstProvider()* method is called to return an address data of the first service provider in the best execution path.

6.5 Message Helper

GoCoMo Message Helper (**GMsgHelper**) is a group of classes that define different GoCoMo messages data and support message generating and parsing. As depicted in Figure 6.8, this work uses the builder pattern to realize **GMsgHelper**. A GoCoMo message consists of header data, content data and extra data. The header data includes information about the message type, request id, the sender's address and the receiver's address. The content data, depending on the message type, can include a composition request or service information. The extra data provides additional content data, like the address of the nearest service provider that hosts a Or-split guidepost.

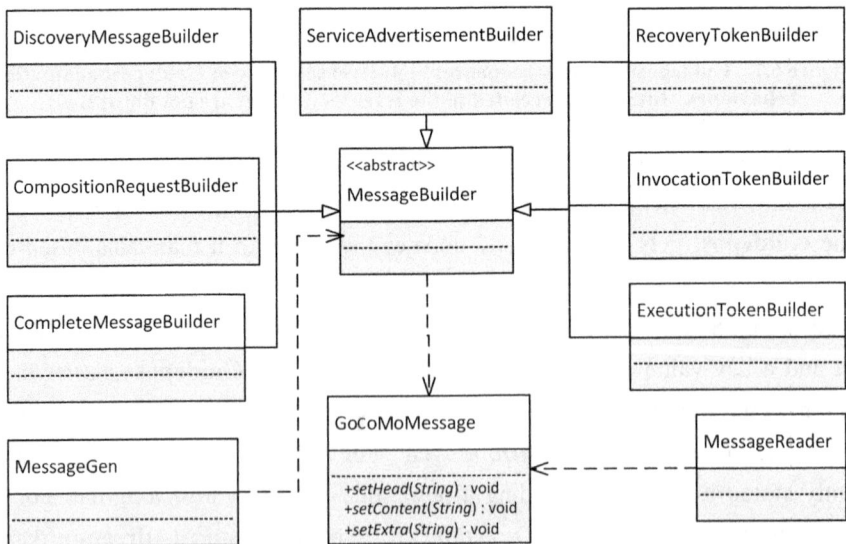

Figure 6.8 GoCoMo messages implementation: class diagram of GoCoMo Message Helper

6.6   Prototypes

Two GoCoMo prototypes were implemented to support a proof of concept. One is a middleware application on Android platforms and the other is an extension module on the ns-3 platform. Both of the proto types realize GoCoMo composition processes, and are designed for the purpose of evaluating GoCoMo's feasibility and performance.

6.6.1   Prototype on Android

As one of the most popular mobile OS, Android has been installed in billions of mobile devices [188-192]. More than 1.5 million applications were available in the Android application market (Google™ Play) by August 2015 [193], covering various categories like education, entertainment, business, health, travel & local, etc. This suggests considerable potential for widespread availability of microservices in pervasive environments. Research on Android-based microservices in pervasive computing environments has led to investigations on cooperating multiple Android mobile devices to support a composed functionality, such as using smart phones and smart bracelets to enable mobile sensing for a health-care environments [194, 195]. The Android OS and Android applications require an effective tool to ease cooperation between devices and microservices.

The GoCoMo middleware prototype implemented for Android-based devices, GoCoMo App for simplicity, contains all the major modules introduced above, and supports GoCoMo composition processes. As for the utility modules in the GoCoMo middleware, GoCoMo App relies on syntactic matchmaking for **Matchmaker**, and realizes the **Data Manager** and **Service Manager**. Implementing semantic matchmaking is out of this work's scope. Figure 6.9 briefly illustrates how GoCoMo App implements the GoCoMo middleware to slot its modules in the Android application framework. As GoCoMo App was designed for the purpose of evaluating GoCoMo in real world, in addition to the GoCoMo middleware, GoCoMo App realizes a visualization module that renders GoCoMo App's real-time information that indicates how a GoCoMo composition process is performed at runtime. To enable a service provisioning network, GoCoMo App employs and implements BlueHoc [196], a bluetooth-based ad hoc network for Android distributed computing. Execution guideposts generated during a GoCoMo composition process are serialized to JSON data and saved using the Shared Preference API provided in Android.

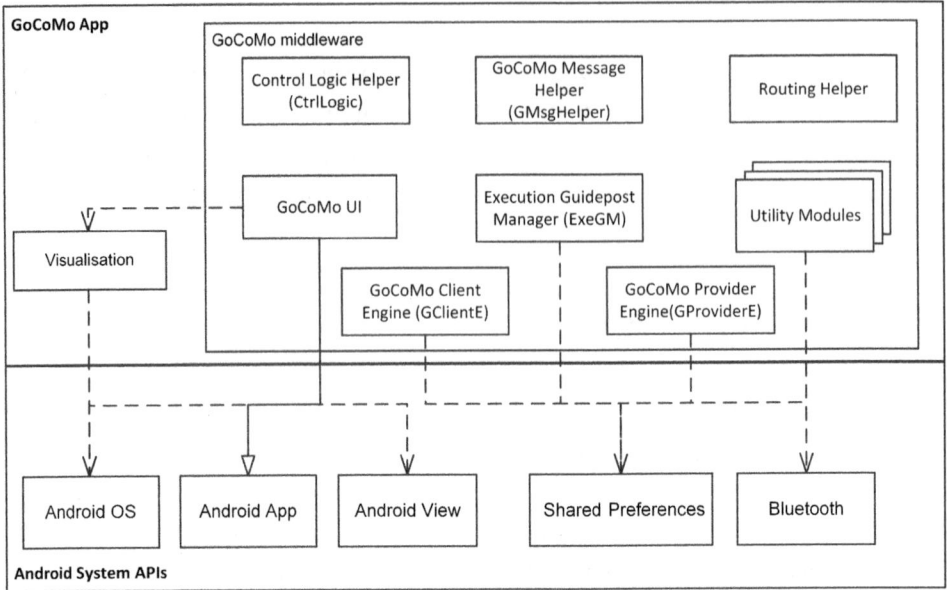

Figure 6.9   GoCoMo prototype on Android: GoCoMo App

More details about GoCoMo App are in Appendix A. Note that GoCoMo App performs the GoCoMo composition process on composition clients' and service providers' devices with an Activity[①] running to display such a process's runtime information. This is of use when evaluating GoCoMo since it allows device owners to monitor their own GoCoMo App's performance. In real world scenarios, keeping a screen active is battery-consuming, which may reduce a device's availability, and so GoCoMo App should be extended to run in the background to reduce energy cost. In addition, GoCoMo App uses a bluetooth ad hoc network.

Bluetooth-based networks have very limited communication ranges, making them a less-competitive technique for wireless communication. A WiFi-based network is preferred because of its large communication range, but enabling WiFi ad hoc requires re-configurations on Android devices, for example, rooting a device or installing a third-party application. GoCoMo App does not go into the above issues as they are out of this work's scope, but these issues should be considered when deploying GoCoMo

---

① An Activity provides a screen with which users can get information or interact in order to modify the local service information or input a composition requirement.

App for real world use.

6.6.2 Prototype on ns-3

Ns-3 is an open-source network simulator based on C++, providing an open, extensible network simulation platform [197]. A ns-3 extension model is a group of classes providing a specific set of functionalities, which includes related classes, examples, as well as tests, and can be used with existing ns-3 modules/models [197]. There's a ns-3 extension model, named GoCoMo-ns3, built on top of ns-3 platforms, which realizes most of the GoCoMo middleware modules. In particular, GoCoMo-ns3 includes all the major modules of the GoCoMo middleware, and a part of the utility modules. The **GoCoMo UI** module and the **Service Manager** module are excluded, because the prototype is designed for evaluating GoCoMo's performance and feasibility in a microservice network, and does not involve actual microservices' execution and user interactions. Figure 6.10 illustrates how the GoCoMo middleware is added to the ns-3 platform as an extension model and how other ns-3 modules/models work with GoCoMo-ns3 to support a simulation of the GoCoMo composition process. Further details in GoCoMo-ns3 are presented in Appendix A, and the experiment configuration will be introduced in the next chapter.

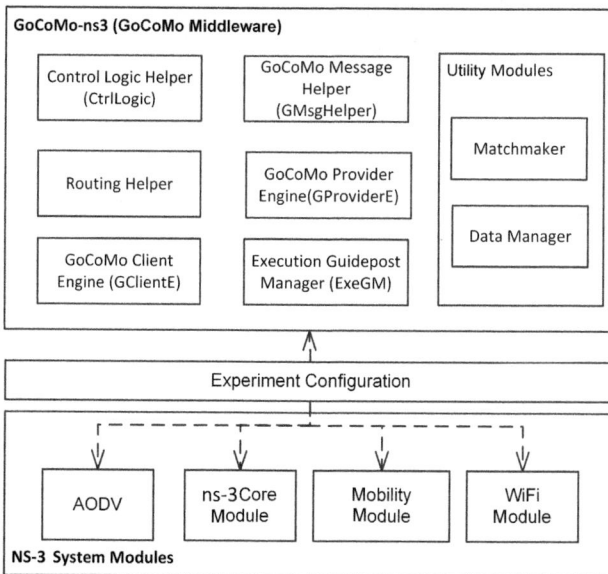

Figure 6.10 GoCoMo prototype on ns-3: GoCoMo-ns3

6.7 Implementation Summary

This chapter introduces the GoCoMo middleware and two prototype implementations on the Android platform and the ns-3 platform. The GoCoMo middleware lies between local microservices and the network on each composite participant.

The GoCoMo composition is mainly realized by a number of major modules, which include **GClienE** and **GProviderE** to coordinate actions across the global composition process and the local composition process, respectively. **ExeGM** and **CtrlLogic** are used to generate a service execution workflow, which may include complex control logic if necessary. In addition, **GMsgHelper** manages all the messages that are used in interactions between composite participants to exchange information. **RoutingHelper** controls such messages' transmission in a network. Utility modules and UI modules underpin GoCoMo composition processes by assisting the major modules to operate matchmaking or obtain context data. Their implementation may differ in varying platforms.

This chapter also describes two prototype implementations for the GoCoMo middleware that are designed for evaluation. The next chapter, based on these prototypes, describes the evaluation of GoCoMo through ns-3-based simulation and a case study in an Android device network, and presents the evaluation results.

Chapter 7

Implementation II: Artificial Intelligence Services

Artificial intelligence has become the core driver of the global technological revolution and industrial change, as well as an important foundation and powerful engine for green, intelligent and sustainable socio-economic development. It studies and develops theories, methods, technologies, and application systems for simulating, extending, and expanding human intelligence. In recent years, the development of AI technology and AI-based applications has become a national strategic policy worldwide. Especially with the breakthroughs in key core technologies of AI in the fields of speech and vision, it has become a common expectation of society that AI technology will further break through the industrial barrier and domain silos, to achieve comprehensive and deep applications.

Edge intelligence, also known as edge-native AI architecture, not only enables end devices to obtain more diverse intelligent services by deploying AI algorithms on computing nodes close to end devices but also significantly reduces the feedback delay of services and energy consumption of end devices, which has become a new trend to realize deep AI applications. Meanwhile, edge intelligence is also a much-needed new component in the current fully commercialized 5G networks and is widely considered as one of the key features of future 6G wireless systems.

Generally speaking, artificial intelligence refers to a computing system with machine learning algorithms as the core functional component, and its related applications are spread across various application areas that have the requirement of processing raw data, such as transportation, logistics, energy, environment, and urban management, etc. The most common solution of a machine learning-based data processing service is to parse data, learn from it, and acquire algorithmic models through iterative training, and then realize decision making or prediction of events in the real world.

This chapter gives a taste of how to implement intelligent services. It includes the implementation of service provisioning frameworks and a step-to-step guideline of AI services deployment. It also provides a running example.

7.1 Service Provisioning Frameworks

Service provisioning frameworks provide tools and libraries to support service provisioning process that may include service registration, service discovery, service composition, QoS monitoring and so on. Since the service composition implementation presented in the last chapter is based on JavaTM, this chapter follows such language setting. The main Java-based microservice development tools currently used by enterprises include Spring Cloud, Dubbo and Dropwizard, etc.

Spring Cloud is a set of microservice solutions based on Spring Boot, which combines mature and proven service frameworks developed by various companies. It re-encapsulates such service frameworks in the Spring Boot style to remove complex configuration and implementation principles, ultimately providing developers with an easy to understand, easy to deploy and easy to maintain distributed system development toolkit. In terms of framework selection, Spring Cloud is easier to understand and use for developers familiar with Spring.

Dubbo is a distributed service governance framework. It was initially open-sourced by Alibaba in 2011 and has formed by the gradual exploration and evolution in Alibaba's e-commerce platform. Its capabilities of dealing with high concurrency challenges of complex business have been proven. Dubbo entered the Apache Incubator on December 15, 2016, renamed as Apache Dubbo. Apache Dubbo is currently used by many leading mobile Internet companies in China, such as Alibaba, JD.com, Dangdang, Ctrip.

Dropwizard integrates stable, mature components of many problem domains in the Java ecosystem into one simple, light weight package, enabling users to quickly build a RESTful service platform and integrate projects to the Dropwizard core. Although there are less than 100 applications reported using Dropwizard, compared with Spring Cloud, Dropwizard's lightweight nature can make it a promising solution for pervasive applications.

This chapter mainly introduces an implementation based on Spring Cloud. The idea of this implementation can also be expanded to other frameworks or languages.

7.1.1   Spring Cloud

Spring Cloud provides developers with a set of tools to quickly build a conventional model for distributed systems including configuration management, service discovery, intelligent routing, micro-proxies, control buses, one-time tokens, global locks, leader elections, distributed sessions, cluster state, etc. This section demonstrates a microservice architecture built on Spring Cloud, which consists of four parts: service registry, service configuration, composition middleware, and service gateway (See Figure 7.1).

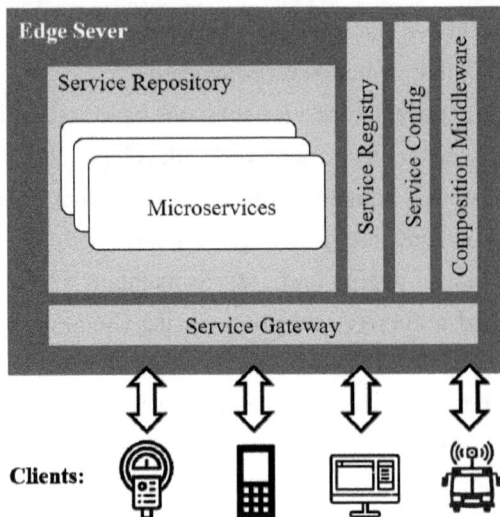

Figure 7.1   Microservice architecture based on Spring Cloud

7.1.2   Service Configuration

To create a microservice environment, Spring Cloud Config Server (SCCS) is employed to support external configuration from both the client-side and the server-side. It provides a centralized portal to configure distributed systems and can easily be embedded in any application based on Spring Boot.

To initialize SCCS, we first create a Maven parent project called "demo-microservice" to manage the versions of project dependency packages. Then we create a sub-project named "config" under "demo microservice" and add Spring Cloud Config Server's dependency package into the "pom.xml" file of the subproject, see below.

```
1. <dependencies>
2.     <dependency>
3.         <groupId>org.springframework.cloud</groupId></groupId>
4.         <artifactId>spring-cloud-config-server</artifactId>
5.     </dependency>
6. </dependencies>
```

Set the location and the server port in the configuration file "application. yml" of Spring Cloud Config Server as follow.

```
1. spring:
2.     application:
```

3.　　　name: config

4.　　profiles:

5.　　　active: native

6.　　cloud:

7.　　　config:

8.　　　server:

9.　　　　native:

10.　　　　　search−locations: classpath: /configs/

11. server:

12.　　port:8000

To run a server for microservices, the SCCS dependency and @EnableConfigServer are used, as shown below in Line 8.

1.　　　package com. cloudbrain.config;

2.

3. import org.springframework.boot.SpringApplication;

4. import org.springframework.boot.autoconfigure.SpringBootApplication;

5. import org.springframework.cloud.config.server.EnableConfigServer;

6.

7. @SpringBootApplication

8. @EnableConfigServer

9. public class ConfigApplication {

10.　　publicstaticvoid main (String [ ] args) {

11.　　　　SpringApplication.run (ConfigApplication.class, args) ;

12.　　}

13. }

7.1.3　Service Registration at Edge

Service discovery is one of the key functions of microservice-based architectures. To provide microservices to a client, microservices deployed in the edge computing network must be able to be found by clients. However, for each edge device, manually configuring each client or using some forms of convention to invoke service can be difficult and fragile. Automatic service discovery mechanisms are required to achieve service management without human involvement.

Zookeeper is an open-source distributed orchestration service under the Hadoop project. Eureka is a service governance module under the Netflix project, which is included in Spring Cloud with Netflix. Eureka allows servers to be configured and deployed in a way of achieving high availability, by replicating state about registered services to different servers. This way makes Eureka's registered services available as long as there is one Eureka server still working, which makes Eureka a suitable service discovery platform in our scenario.

A four-step configuration can be used to enable Eureka Server.

Step 1: create a new sub-project "registry" under "demo-microservice" and add Eureka Server dependencies to the "pom.xml" file under this project, as shown below.

```
1. <dependencies>
2.    <dependency>
3.        <groupId>org.springframework.cloud</groupId>
4.        <artifactId>spring-cloud-netflix-eureka-server</artifactId>
5.    </dependency>
6.    <dependency>
7.        <groupId>org.springframework.cloud</groupId>
8.        <artifactId>spring-cloud-starter-config</artifactId>
9.    </dependency>
10. </dependencies>
```

Step 2: create a directory named "configs" to the resources of the configuration center project, and add the configuration file "registry.yml" with the following configuration.

```
1. spring:
2.    application:
3.        name: registry
4.
5. eureka:
6.    client:
7.        register-with-eureka:false
8.        fetch-registry: false
9.        serviceUrl:
10.           defaultZone: http://localhost: ${server.port}/eureka/
11.
12. server:
13.    port:8001
```

Step 3: create a configuration file "bootstrap.yml" to the resource of registry and add the following information. Note that the port number can be different according to the deploy environment.

```
1. spring:
2.    cloud:
3.       config:
4.          name: registry
5.          uri: http: //localhost:8000
```

Step 4: add @EnableEurekaServer to enable the service registry.

```
1. package com.cloudbrain.registry;
2.
3. import org.springframework.boot.SpringApplication;
4. import org.springframework.boot.autoconfigure.SpringBootApplication;
5. import org.springframework.cloud.netflix.eureka.server.EnableEurekaServer;
6.
7. @SpringBootApplication
8. @EnableEurekaServer
9. public class RegistryApplication {
10.       publicstatic void main (String [ ] args) {
11.                SpringApplication.run (RegistryApplication.class,args);
12.       }
13. }
```

7.1.4   Service Gateway

Service gateway allows clients to locate service providers in the network.
In our case, service gateway can be configured through Eureka Client.
To do so, we first create a new sub-project "gateway" under "demo-microservice" and add the following dependency packages.

```
1. <dependencies>
2.       <!-- zuul -->
3.       <dependency>
4.          <groupId>org.springframework.cloud</groupId>
5.          <artifactId>spring-cloud-starter-netflix-zuul</artifactId>
6.       </dependency>
```

```
7.
8.      <!-- EurekaClientStarter -->
9.      <dependency>
10.        <groupId>org.springframework.cloud</groupId>
11.        <artifactId>spring-cloud-starter-netflix-eureka-client </artifactId>
12.      </dependency>
13.
14.      <!-- ConfigClientStarter -->
15.      <dependency>
16.        <groupId>org.springframework.cloud</groupId>
17.        <artifactId>spring-cloud-starter-config</artifactId>
18.      </dependency>
19. </dependencies>
```

Then, add the configuration file "gateway.yml" to the "configs" directory under the project resources in the configuration center, with the following configuration.

```
1. spring:
2.    application:
3.      name: gateway
4.
5. Server:
6.    port:8002
7.
8.
9. eureka:
10.    client:
11.      serviceUrl:
12.        defaultZone: http://localhost:8001/eureka/
13.    instance:
14.      lease-renewal-interval-in-seconds: 10
15.      lease-expiration-duration-in-seconds: 60
16.      prefer-ip-address: true
17.      instance-id: ${spring.application.name}: ${spring.application.instanceid}:
         ${server.port}}
```

Add "bootstrap.yml" to the gateway's resources and configure it as follows.

```
1. spring:
```

2. cloud:

3.   config:

4.   name: gateway

5.   uri: http://localhost:8000

Create startup class and add comment @EnableZuulProxy to enable gateway proxy service.

1. packagecom.cloudbrain.gateway;

2.

3. importorg.springframework.boot.SpringApplication;

4. importorg.springframework.boot.autoconfigure.Spring BootApplication;

5. importorg.springframework.cloud.netflix.zuul. Enable ZuulProxy;

6.

7. @SpringBootApplication

8. @EnableZuulProxy

9. public class GatewayApplication {

10.   publicstatic void main (String [ ] args) {

11.   SpringApplication.run (GatewayApplication.class,args) ;

12.   }

13. }

7.2   Deploy AI Models

This section briefly introduces a way to deploy a trained deep learning algorithm to our systems as a microservice and the way of accessing the microservice through the service provisioning framework.

Applying a deep learning algorithm usually includes two steps: model training and model deployment. The training of a supervised deep learning model is to input a large number of training samples and corresponding labels into the model and continuously update the parameters until the model finally has the ability to map the input data to the corresponding labels. The model deployment is the process of making the model fit in the execution environment after the completion of model training, so the model can take raw data as input and generate corresponding result as the output.

The flexibility and complete ecological library of Python make it a perfect choice for implementing and verifying deep learning algorithms. In fact, it has become a dominant programming language in deep learning area with the release of open-source Python libraries such as PyTorch and Tensorflow. This section will introduce three solutions to deploy algorithms written in Python into the Java-based microservice architecture built in the previous sections.

1. A commonly used solution is to compile a piece of Python code and pack it into a JAR (Java Archive) file so that the Python code can be called directly by Java through jython. But this way is limited by the capabilities of jython which only supports the earlier versions of Python, such as Python 2.7. In fact, many state of the art AI algorithms have already shifted to Python 3. In addition, this method does not support packaging for many third-party libraries, such as NumPy.

2. The Python-based model runs in the Python environment through Flask. A microservice platform can access the model through an HTTP protocol with Flask. Java is a statically typed language, which is compiled in advance and executed the bytecode directly, while Python is a dynamically typed language, which is compiled while executing, so the compilation of Python is slower than Java. The initialization of a Python-based model can take a while. For better performance, it is not recommended for applications that require efficient AI model invocation using the Python environment.

3. TensorFlow provides a Java API for loading models created in Python and running them in Java-based applications. It can run on any JVM (Java Virtual Machine) and can be used to build, train, and deploy machine learning models.

This section demonstrates a solution implemented using the third solution for Python-based deep learning algorithm model deployment.

The configuration introduced in Section 7.1 forms a microservice architecture foundation which allows microservices to be deployed. This section takes a simple deep learning task as an example to illustrate how to deploy it into an edge device as a microservice. The following content of this section will describe the ways of offline model packing, model deployment, as well as frameworks to train an AI model.

7.2.1 Packed as a Microservice

Image recognition is a typical and widely used computer vision task that is based on AI algorithm to recognize objects in an image. This section applies the MNIST database to train an image recognition algorithm in Python with Tensorflow to recognize handwritten digits.

To prepare a microservice, we need to wrap-up the algorithm. After the model is trained in Python, save it as a .pb file. The content of the input data, output data and tags are required to record, because those three items will be used when the model is loaded. The saved. pb file contains not only the weights of the parameters, but also the computational graph, which is language-independent, run-independent, and in a closed serialized format that can be parsed by any language, allowing other languages and deep learning frameworks to read, continue training, and migrate the algorithm. The file structure is shown in Figure 7.2.

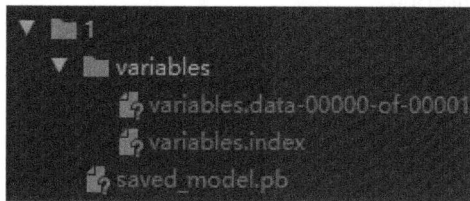

Figure 7.2   File structure of the model in Python

To deploy the algorithm in a Java-based microservice framework, the TensorFlow Java dependency package needs to be added into pom.xml, as follows.

```
1. <dependency>
2.        <groupId>org.tensorflow</groupId>
3.              <artifactId>tensorflow</artifactId>
4.        <version>1.14.0</version>
5. </dependency>
```

After that, a controller layer should be added to the MNISTPredict algorithm module to connect the client with the server that hosts the microservice. The controller receives a service request and calls the corresponding microservice through a service layer. An implementation of the controller layer (i.e., MNISTController) and the service layer (i.e., MNISTService) are shown as follows.

```
1. package com.mnistpredict.controller;
2.
3. import com.mnistpredict.service.MNISTService;
4. import org.springframework.beans.factory.annotation.Autowired;
5. import org.springframework.web.bind.annotation.GetMapping;
6. import org.springframework.web.bind.annotation.RequestMapping;
7. import org.springframework.web.bind.annotation.ResponseBody;
```

8. import org.springframework.web.bind.annotation.RestController;

9.

10. @RestController

11. @RequestMapping ("/")

12. public class MNISTController {

13.　　　　@Autowired

14.　　　　MNISTService mnistService;

15.

16.　　　　@GetMapping ("/mnist")

17.　　　　@ResponseBody

18.　　　　public String predict ( ) {

19.　　　　　　return mnistService.MNISTPredict ( );

20.　　　　}

21. }

22. package com.cloudbrain.mnistpredict.service;

23.

24. public interface MNISTService {

25.　String MNISTPredict ( ) ;

26. }

The service layer invokes a service implementation layer to initial the MNIST service. The algorithm model calls are written in the service implementation layer. A realization of the service implementation layer (i.e., MNISTServiceImpl) is shown as follows. Note that the method MNISTPredic () defined in the service layer is rewritten in the service implementation layer.

1. package com.cloudbrain.mnistpredict.service.Impl;

2.

3. import com.cloudbrain.mnistpredict.service.MNISTService;

4. import com.cloudbrain.mnistpredict.utils.ImageProcess;

5. import org.springframework.stereotype.Service;

6. import org.tensorflow.SavedModelBundle;

7. import org.tensorflow.Session;

8. import org.tensorflow.Tensor;

9.

10. import javax.imageio.ImageIO;

11. import java.awt.image.BufferedImage;

12. import java.io.File;

13. import java.io.IOException;

14. import java.util.Arrays;

15.

16. @Service

17. public class MNISTServiceImpl implements MNISTService
        {

18.   SavedModelBundle tensorflowModelBundle;

19.   Session tensorflowSession;

20.

21.   void load (String modelPath){

22.   this.tensorflowModelBundle = SavedModelBundle.load (modelPath ,
     "serve");

23.   this.tensorflowSession = tensorflowModelBundle.session ( );

24.   }

25.

26. public Tensor predict (Tensor tensorInput){

27.   Tensor output = this.tensorflowSession.runner ( ).feed ("x:0", ten-
     sorInput).fetch ("y:0").run ( ).get (0);

28.   return output;

29.   }

30.

31.   @Override

32.   public String MNISTPredict ( ) {

33.   ImageProcess imageProcess = new ImageProcess ( );

34.   BufferedImage bf = imageProcess.readImage ("G: \\ images \\ 7.jpg");

35.   float [ ] [ ] testvec = imageProcess.convertImageToArray (bf);

36.   Tensor input = Tensor.create (testvec);

37.

38.   MNISTServiceImpl myModel = new MNISTServiceImpl ( );

39.   String modelPath = "G: \\ 1 \\ ";

40.   myModel. load (modelPath);

41.

42.   Tensor out = myModel.predict (input);

43.   float [ ] [ ] resultValues = (float [ ] [ ]) out.copyTo (new float [1] [1 0]);

44.　　　　input.close ( );

45.　　　　out.close ( );

46.　　　　return "the result is: " + Arrays.toString (resultValues [0]);

47.　　}

48. }

In MNISTServiceImpl, lines 22 and 27 take care of the recorded contents when the model is saved, and lines 40 and 42 are for model loading and model prediction. A specific implementation of the image processing model is illustrated as follows.

1. package com.cloudbrain.mnistpredict.utils;

2.

3. import javax.imageio.ImageIO;

4. importjava.awt.image.BufferedImage;

5. importjava.io.File;

6. importjava.io.IOException;

7.

8. public class ImageProcess {

9.　　public static BufferedImage readImage (String imageFile) {

10.　　File file = new File (imageFile);

11.　　BufferedImage bf = null;

12.　　try {

13.　　　bf = ImageIO.read (file);

14.　　} catch (IOExceptione) {

15.　　　e.printStackTrace ( );

16.　　}

17.　　return bf;

18.　　}

19.

20. public static float [ ] [ ] convertImageToArray (BufferedImage bf) {

21.　　int width = bf.getWidth ( );

22.　　int height = bf.getHeight ( );

23.　　int [ ] data = new int [width*height];

24.　　bf.getRGB(0, 0, width, height, data, 0, width);

25.　　float [ ] [ ] rgbArray = new float [height] [width];

26.　　for (int i =0; i<height; i++)

```
27.        for (int j = 0; j<width; j++)
28.            rgbArray [i] [j] = data [i*width +j] ;
29.    return rgbArray;
30.    }
31. }
```

7.2.2 Microservice Deployment

To deploy a microservice, we create a new sub-project "MINISTpredict" under "demo-microservice" to recognize digital images in the handwritten dataset MNIST using a deep learning model and add the following dependency packages into file pom.xml.

```
1. <dependencies>
2.      <!-- Spring Boot Web Starter -->
3.      <dependency>
4.          <groupId>org.springframework.boot</groupId>
5.          <artifactId>spring-boot-starter-web</artifactId>
6.      </dependency>
7.
8.      <!-- feign -->
9.      <dependency>
10.         <groupId>org. springframework.cloud</groupId>
11.         <artifactId>spring-cloud-starter-openfeign</artifactId>
12.     </dependency>
13.
14.     <!-- Eureka Client Starter -->
15.     <dependency>
16.         <groupId>org.springframework.cloud</groupId>
17.         <artifactId >spring-cloud-starter-netflix-eureka-client</artifactId>
18.     </dependency>
19.
20.     <!-- Config Client Starter-->
21.     <dependency>
22.         <groupId>org.springframework.cloud</groupId>
23.         <artifactId>spring-cloud-starter-config</artifactId>
```

24.   </dependency>

25.

26.   <dependency>

27.   <groupId>org.tensorflow</groupId>

28.   <artifactId >tensorflow</ artifactId >

29.   <version>1.14.0</ version>

30.   </dependency>

31. </dependencies>

In the configuration center, add the configuration file "mnistpredict.yml" to the "configs" directory of the project resources, with the following configuration.

1. spring:

2.   application:

3.   name: mnistpredict

4.

5. server:

6.   port: 8010

7.

8. eureka :

9.   client:

10.   serviceUrl:

11.   defaultZone: http: //localhost: 8001/eureka/

12. instance:

13.   lease−renewal−interval-in−seconds: 10

14.   lease−expiration−duration−in−seconds: 60

15.   prefer−ip−address: true

16.   instance−id: ${ spring.application.name } : ${ spring.application.instanceid} ${ server.port }}

Add "bootstrap.yml" to the "mnistpredict" resource and configure it as follows.

1. spring:

2.   cloud:

3.   config:

4.   name: mnistpredict

5.   uri: http: //localhost: 8000

Create a startup class and add @EnableEurekaClient to enable service registration and service discovery for this microservice. After that, add @EnableFeignClients to enable Spring Cloud Feign support.

```
1. package com.mnistpredict;
2.
3. import org.springframework.boot.SpringApplication;
4. import org.springframework.boot.autoconfigure.SpringBootApplication;
5. import org.springframework.cloud.netflix.eureka.EnableEurekaClient;
6. import org.springframework.cloud.openfeign.EnableFeignClients;
7. import org.springframework.web.bind.annotation.GetMapping;
8. import org.springframework.web.bind.annotation.RequestMapping;
9.
10. @SpringBootApplication
11. @EnableEurekaClient
12. @EnableFeignClients
13. public class MNISTApplication {
14.    public static void main (String [ ] args) {
15.            SpringApplication.run ( MNISTApplication.class,args) ;
16.    }
17. }
```

So far the basic modules in the microservice framework have been built, including the service registry, the service configuration center, the service cluster, and the service gateway. A microservice can be started by successively running the following module's startup class: ConfigApplication, RegistryApplication, GatewayApplication and MNISTApplication.

The above steps achieve the whole process of deploying Tensorflow deep learning algorithm models to Java microservices. Users can access the MINISTPredict algorithm service through the defined gateway (port:8002) and input a handwritten digital image of size 28 × 28 (as shown in Figure 7.3)to the model. The output of this microservice is shown in Figure 7.4.

Figure 7.3   A handwritten digital image as input

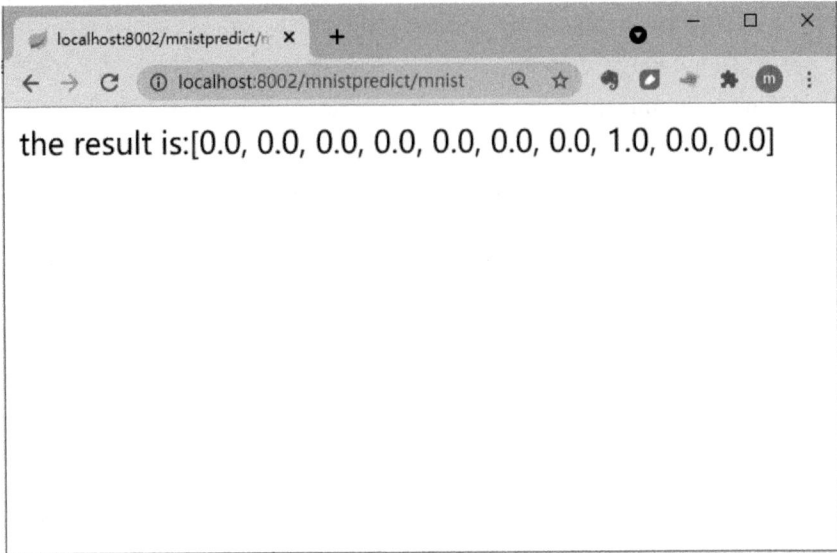

Figure 7.4   Output of the MNIST service

The display shows the predicted result for the number 7 corresponding to the input data.

7.2.3   Platforms for AI Services

The last section demonstrates the microservice deployment with Tensorflow. This section introduces two other platforms that also support the service provisioning framework.

Keras is a deep learning library based on TensorFlow and Theano. It is a high-level neural network API written in pure Python and supports Python development only. There are two ways to deploy Keras models as a microservice in the demonstrated framework.

1. Convert Keras' hierarchical data format model to Tensorflow's Protocol Buffers model (.pb), and then deploy the model using the way described in Section 7.2.2.
2. Use Deeplearning4j that is a global commercial-grade open-source deep learning library launched by Skymind for Java users. It supports importing Keras-trained models and provides functions similar to NumPy in Python to handle structured data. However, to the best of our knowledge, Deeplearning4j only covers the versions prior to Keras 2.0.

PyTorch is an open-source Python-based machine learning library based on Torch.

It is easy to use and enables not only powerful GPU acceleration, but also supports dynamic neural networks, so there's an increasing number of deep learning models built with PyTorch. There are two commonly used ways to deploy PyTorch-based models in the demonstrated framework.

1. Convert the PyTorch-based model to a Tensorflow-based model first, then deploy it as a microservice using the method mentioned in Section 7.2.2.

2. Use Deep Java Library (DJL). DJL is a deep learning platform built entirely in Java, which allows deep learning model development in Java and supports the PyTorch framework. It allows TorchScript format models that contains the model structure and parameters to be deployed in the platform. Models in other formats need to be converted to TorchScript before deployment using the *torch.jit.trace*() that captures the model architecture.

7.3   Challenges for AI–based Services Composition

The application of machine learning algorithms is generally based on the assumption of homogeneous distribution between the training data and the data of real application scenarios, which means that the difference in data distribution between them will directly affect the data processing quality of the algorithm. For example, object detection algorithms trained only in bright scenes have poor detection accuracy in dark scenes, or in the rain, snow, or fog scenes. Autonomous driving algorithms trained in closed campus roads are also difficult to adapt to the demand for autonomous driving in complex urban road conditions.

7.3.1   Feature Heterogeneity

With the development of AI technology, it becomes new trends that relying on edge computing to train machine learning algorithms, and deploying machine learning algorithms in edge networks to provide intelligent data processing services for end users. However, due to the different data sampling environments of end devices, the diversity of data sources in terms of their distribution feature exists. The limited computational resources of edge nodes can hardly carry an AI algorithm adapted to all data distributions at the same time, thus requiring the deployment of AI algorithms that meet the real data distribution requirements. It is time-consuming and labour-intensive to design a deployment scheme for each edge node to meet every dynamic data requirement.

7.3.2 High-dimensional Data

In the main application areas of machine learning (e.g., machine vision, natural language processing), there is a large amount of high-dimensional data, such as images/video, speech, etc. The feature distribution of these high-dimensional data is difficult to be represented by simple and intuitive modelling methods. Therefore, for wide-area coverage of edge intelligence, it is important to establish the mapping relationship between real data distribution and machine learning algorithms, so as to efficiently deploy adapted AI algorithms in edge computing networks.

7.3.3 Dynamic Raw Data

The edge computing environment is full of dynamic characteristics. The data collected by terminal devices may be affected by various environmental dynamic factors such as weather, light, background noise, etc., and many factors such as terminal movement and data volume fluctuation may also cause changes in data distribution. It is the fact that the developing AI technology will gradually improve the algorithm generalization ability, making the algorithm itself powerful enough to adapt to different data distribution scenarios. However, even with a powerful algorithm, the training data acquisition and labelling are still expensive. The enhanced algorithm capability also means higher computing power requirements, which are difficult to be met by resource-constrained edge nodes. Therefore, it is still a challenge to achieve online deployment of AI algorithms in dynamic computing environments to ensure real-time, and to reasonably schedule data processing tasks to avoid service failure in response to data volume fluctuations.

7.4 Implementation Summary

This chapter provides a practical and detailed example of deploying a Tensorflow-based AI service in a Java-based computing environment. It includes algorithm training, service warping (a.k.a., model packing), service discovery gateway, service deployment and invocations. Step-by-step implementation and configurations are also provided in this chapter. As AI services are normally developed and trained in Python with different platforms, this chapter gives an introduction on three ways to enable AI services considering the differences on their training platforms and introduces model conversion methods for two popular platforms: PyTorch and Keras. This chapter also analyzes the remaining challenges for AI services provisioning in edge computing environments.

Chapter 8

Evaluation

THE previous chapter introduces the GoCoMo middleware and the details of its implementation, which underpins the GoCoMo composition process proposed in Chapter 3. This chapter evaluates how well the GoCoMo approach addresses the target environment by satisfying the required features specified in Section 3.1. The four evaluation objectives below are realized in this chapter.

1. To determine whether the GoCoMo middleware can flexibly reason about a service composite at runtime, in a decentralized manner, and self-organize the composite's execution and adaptation.

2. To quantify the success of GoCoMo's service planning, heuristic request routing and composition adaptation model, in terms of planning and execution failure probability.

3. To evaluate the performance of GoCoMo, determining its feasibility in a range of mobile and pervasive networks/scenarios.

4. To evaluate the performance of LOCASS in service provisioning network with social selfishness.

This chapter first outlines the evaluation method and a list of evaluation criteria, and then introduces a case study that achieved evaluation objective-1. After that, a simulation is used to address evaluation objective-2 and evaluation objective-3. The GoCoMo composition process is examined under different scenarios in mobile and pervasive environments. The validity of the evaluation results is discussed in Appendix B.

8.1   Evaluation Methods and Criteria

Given the challenges outlined in this book, a usable service composition model for pervasive computing should be able to tackle infrastructure-less networks, providing a dynamic, decentralized service composition process. The service composition process itself should be sufficiently time-efficient and adaptable to reduce composition failures caused by changes to the network and the service topology. This chapter uses a simulation study and a prototype case study to evaluate GoCoMo.

A case study is an observation-based method that investigates a single entity's activity within a specific time space [198-200], and can be used to evaluate the benefits of methods and tools [201]. Prototype case studies relying on real-world pilot implementations can reflect the behaviour of a model in the real world scenario [202], based on which, the model's feasibility, reliability or flexibility can be investigated [199]. Many service compositioning models and composite adaptation strategies proposed to support service provisioning have adopted prototype case studies [203-207].

This chapter builds a testbed that relies on Android platforms, and deploys a pilot implementation of GoCoMo on the testbed to conduct a prototype case study. Testbeds, sometimes called experimentation networks, are in-lab networks established and used by researchers [208]. In general, testbeds have limited scalability in terms of network size because of the cost of hardware and the monitoring/deployment difficulties, but capture more aspects that influence the performance of algorithms and protocols comparing to software-based simulators [202]. The prototype case study focuses on the following evaluation metrics.

1. Support for pervasive environments where previously cached conceptual composites are impossible (Challenge 1).
2. Support for infrastructure-less networks and fully decentralized service discovery and execution (Challenge 2).
3. Support for service composite adaptation when the operating environment is dynamic and open (Challenge 3, 4 and 5).

In the domain of decentralized service composition that assumes multiple composition planners, the network will include a number of real devices that have enough resources to perform a local planning process. Using a number of real devices is always impractical in laboratory-scale evaluation studies, making it difficult to demonstrate a model's performance in a scalable network (i.e., where the number of nodes is equal to or above 20). Many researchers in the pervasive computing domain, especially those have particular focus on mobile ad hoc networks, consider simulation studies as a helpful evaluation method. The majority of them use only simulations, and several others combine a small scale (using 5-25 nodes) prototype case study and a simulation study to evaluate their approach [209].

Simulation studies that take a network simulation as a controlled experiment have been widely used in service-oriented computing research [56, 210]. A simulation-based experiment models an algorithm with a high degree of abstraction and performs it in an artificial software environment, which makes the experiment effective, repeatable, controllable, and scalable [202, 211]. Such an experiment relies on the simplification of real-world scenarios, manipulating a series of environmental or systemic variables to get an algorithm's performance values in diverse scenarios or under different system configurations. [199, 200].

This chapter quantifies GoCoMo's composition failures and performance, to demonstrate GoCoMo's advantages in a set of scenarios through a simulation-based experiment. The experiment used the following evaluation criteria.

1. Planning failure rate: GoCoMo's success at finding sequential solutions to a goal request. The service composition planning failure rate is calculated as the ratio ($\in$ [0, 1]) of the number of failed planning processes to the number of all the issued requests during the simulation cycles. The duration for a client to receive the first pre-execution plan and the sent messages (system traffic) during this process were also measured for performance analysis.

2. Execution failure rate: GoCoMo's success at handling potential failures that may appear during service execution. In this experiment, the execution failure rate is computed as the ratio ($\in$ [0, 1]) of the number of failed executions to the number of all the successful planning processes, considering different system configurations (i.e., mobility, and network size). This experiment also included measurements for execution performance, such as response time for a client to receive the execution result and the system traffic during this process.

3. The failure rate for composing parallel service flows: GoCoMo's success at finding and invoking parallel solutions. Composition failure rates were calculated under different service availability configurations.

8.2 Prototype Case Study

The prototype case study deployed the GoCoMo middleware in a real world pervasive network, and discovered the feasibility of GoCoMo's decentralized backward planning algorithm and adaptation mechanism. The Android-based prototype implementation is named GoCoMo App and was introduced in Section 6.6.1 and Appendix A.1.

Testbeds have been used in the Mobile ad hoc Network (MANET) domain, and usually realize a network containing less than 50 nodes [208]. With the increasing number of advanced but expensive mobile devices to be used to support pervasive environments, many researchers have built testbeds on a network involving only a small number (< 10) of such devices [212, 213] to evaluate algorithms or protocols in MANETs.

This case study developed a testbed with 8 mobile Android devices marked as Device 0 to 7. Table 8.1 presents the basic information about the devices and their configurations. The GoCoMo App was deployed on the Android OS with API level from 19 to 22, running across a range of mobile devices. The devices have diverse computing power ranging from a CPU speed of 2 GHz and 1 GB RAM to a Quad-core 2.5 GHz CPU and 3 GB RAM, which covered most mainstream manufacturer brands and best-selling Android devices then. The GoCoMo App was developed to visualize the GoCoMo composition process at runtime, and the dynamic system performance on each participated device was monitored by an Android Device Monitor embedded in Android Studio IDE. Figure 8.1 illustrates all the devices used in the study and the testbed system in operation.

Table 8.1   Devices used in case studies (M.M: Manufacturer and Model Number, OS: Android OS version)

Device	OS	API	M.M	CPU	RAM
0	5.1.1	22	Nexus 7 (2013)	Quad-core 1.5 GHz	2 GB
1	5.1.1	22	Nexus 7 (2013)	Quad-core 1.5 GHz	2 GB
2	4.4.2	19	SM-N9006 (Galaxy Note 3)	Quad-core 2.3 GHz	3 GB
3	5.0.2	21	Motorola G-2	Quad-core 1.2 GHz	1 GB
4	4.4.2	19	SM-N9005 (Galaxy Note3)	Quad-core 2.3 GHz	3 GB
5	5.1.1	22	OnePlus One	Quad-core 2.5 GHz	3 GB
6	4.1.2	16	Motorola RAZR i	2 GHz	1 GB
7	5.1.1	22	SM-G920F (Galaxy S6)	Quad-core 2.1 GHz	3 GB

Figure 8.1   All the devices used in the study, and the testbed system in running

8.2.1   Case Study Configurations

This case study adopted 8 devices, 7 of them (device 1-7 in Table 8.1) were service providers and one (device 0 in Table 8.1) was the composition client. They all installed the same version of the GoCoMo App. These devices were connected to a desktop that ran the Android Device Monitor. To demonstrate the feasibility of Go-CoMo, a set of scenarios were used, varying in composition complexity and service availability. The scenarios used in the study are described in Table 8.2.

The composition complexity was determined by microservices available in a network. Similar to those used in [35, 122], the case study adopted microservices that convert alphabets. More specifically, a microservice $S_{A \to B}$ can produce an output of type B if it receives an input of type A [35], and is named *linear-service*. A microservice $S_{DF\leftrightarrow Z}$ can produce an output of type Z if it receives two inputs of type D and F [122], and is named *join-service*. The case study used 8 unique linear-services and 1 join-service in 8 different scenarios (See Table 8.2), in each of which a subset of these microservices were deployed on the 7 service providers. Every microservice subset supports a particular service flow to transfer data A to Z. The service flows for the 8 scenarios are shown in Figure 8.2. An individual service provider hosts only one microservice, and a copy of a microservice can be deployed on one or more providers. The case study achieved diverse composition complexities through different service flows for the same composition goal, which was for microservices that can support alphabet transformations from A to Z. This case study used 4 service flows varying in composition length (the number of service instances per request).

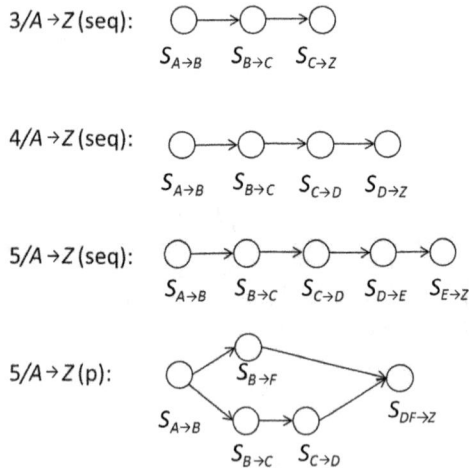

Figure 8.2 Service scenarios used in the case study

Table 8.2 Scenarios for the case study: (seq) = Sequential Service Flow, (p) = Parallel Service Flow, P = Probability of *Wake* State, D_d = Duration of *Wake/Sleep* State

Parameter	Configuration	Scenario							
		1.1	1.2	1.3	1.4	1.5	1.6	1.7	1.8
Microservices	$S_{A\to B}$	★	★	★	★	★	★	★	★
	$S_{B\to C}$	★	★	★	★	★	★	★	★
	$S_{C\to D}$		★	★	★		★	★	★
	$S_{D\to E}$			★				★	
	$S_{C\to Z}$	★				★			
	$S_{D\to Z}$		★				★		
	$S_{E\to Z}$			★				★	
	$S_{B\to F}$				★				★
	$S_{DF\to Z}$				★				★

Continued

Parameter	Configuration	Scenario							
		1.1	1.2	1.3	1.4	1.5	1.6	1.7	1.8
Service flows (service instances per request /composition goal)	$3/A \to Z$ (seq)	★	★			★			
	$4/A \to Z$ (seq)						★		
	$5/A \to Z$ (seq)			★				★	
	$5/A \to Z$ (p)				★				★
Microservice availability	$P = 1$ (static)	★	★	★	★				
	$P = 0.8, D_d = 1$ s					★	★	★	★

Service providers' mobility is one of the most important factors that causes changes in microservice availability. However, mobility models are difficult to apply in a real-world implementation. The case study used a Wake/Sleep pattern for service providers to manage their availability. Every service provider has two states: Wake and Sleep, and the default state is Wake. During the Wake state, the service provider can process GoCoMo messages and service requests. When the Sleep state gets activated, the service provider freezes any composition process it participates in, and stops its communication with any other entities in the network. In other words, during the Sleep state, the service provider is inaccessible, but still keeps the cached data of composition processes. Each state's activation and duration are determined only by the local device, according to a predefined probability and random variables. As presented in Table 8.2, this case study used a static network that keeps all service provider wake and a dynamic network that allows service providers to periodically ($D_d = 1$ s) "flip a coin" to decide if they remain to their current state or activate the other state. The possibility for a "Wake state" is 0.8 ($P = 0.8$).

In the above scenarios, no centralized knowledge base or central composition controller was used. In particular, when a GoCoMo system gets initialized, each service provider only has information about its own microservice, which is the only knowledge that can be used at the start of a local service composition process. Such knowledge is expanded through interactions with other participants during composition processes. For example, by joining a testbed network a participant can obtain a list of addresses of its direct neighbours, and by receiving a GoCoMo discovery message, a participant can get the path length of discovered composites.

8.2.2 Samples and Results

The case study evaluation measured the performance of GoCoMo from each individual mobile device's point of view. These measurements contain observations of two service composition cases. The composition planning case adopted Scenario 1.1-1.4, using evaluation metric 1 and 2 to assess whether the proposed backward planning algorithm can support complex composition planning in infrastructure-less networks. The number of failed planning-based discoveries out of 50 composition attempts were counted in each experiment. The adaptation case applied evaluation metric 3 in Scenario 1.5-1.8, comparing against a service composition model with no adaptation support, to assess whether the proposed adaptation mechanism can reduce execution failure in a dynamic environment. The case study also assessed GoCoMo's maximal CPU usage and memory usage on different mobile devices.

1. Composition Planning Case

GoCoMo's feasibility in a network that is infrastructure-less and requires flexible composition planning were demonstrated from the perspectives of both composition clients and service providers. A composition client measured the overall discovery and execution failures as well as their response time, and the service providers that participated in a composition process measured the time spent on local composition planning and execution.

The composition client initialized a time $T_{discovery}$ for the composition discovery process after a composition request was sent, and the client detected and recorded a discovery failure if no composition complete token has been received when $T_{discovery}$ expired (See Section 4.2). Execution failures were measured in a similar way and used an execution time T_{exe}. The response time for service discovery was the duration from the time a client issues a composition request to the first discovery result being returned to the client, and the response time for service execution was the duration from the time a composition invocation token is sent to receive a composition result.

Figure 8.3(a) illustrates the number of GoCoMo's discovery and execution failures out of 50 attempts on a static network, and (b) shows the response time. The results show that GoCoMo was able to compose and execute different service flows depending on existing microservices using the same composition goal. It was difficult to avoid failures given the use of unreliable bluetooth-based communication channels and limited number of service providers. The response time was decoupled, and represented by the discovery time as the green parts shown in Figure 8.3(b) and the execution time (the grey parts). The discovery time was in the range of 9.241-12.248 s. It was much beyond the tolerable value for a simple information query of 2 s[①]. However, this range is around the upper limit of mobile users' acceptable waiting time for applications, approximately 7-12 s [214]. Similarly, the length of execution was in the range of 7.142-11.992 s, and did not exceed the tolerable value for a task operation, 15 s[②].

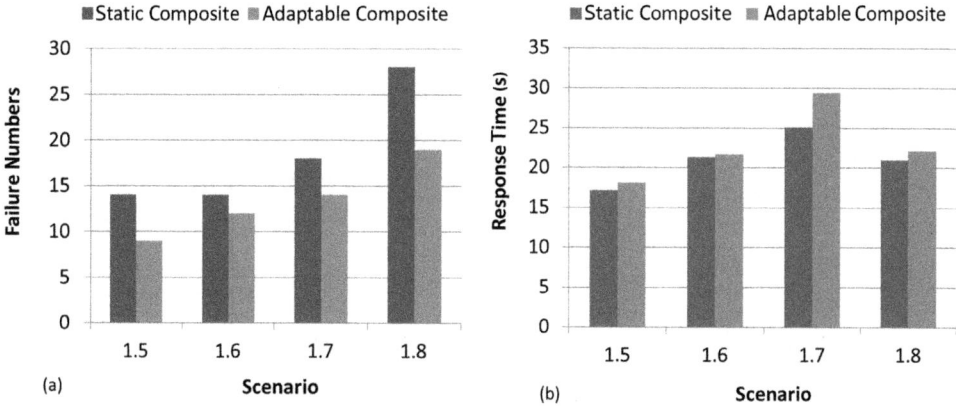

Figure 8.3 GoCoMo's feasibility on static networks
(a) discovery and execution failures out of 50 attempts on a static network, (b) response time for the composition discovery process and the service execution process

Table 8.3 illustrates the performance of GoCoMo on each individual service provider. The build time was the time spent on initializing the GoCoMo App, which varies depending on the deployment device's computation power. The planning time was the duration of a local composition planning process starting from receiving a composition request and ending when a fragment of an execution path (a direction in the guidepost, see Section 6.4.1) is generated. A device started timing the execution process right after receiving a service invocation message, and stopped before sending out a message to invoke the next microservice. Msg represents the number of messages generated and transmitted from the device during composition, and Msg size represents the average size of these messages. The results show that the local composition process itself is efficient as the average performance time (time spent on planning and execution) for every device is small (< 200 ms). As the evaluation network was established in ad hoc mode, establishing message transmission channel was slow, which causes the high global response time, as shown in Figure 8.3(b). Service providers only sent 2 messages, and the messages' size were 227 (byte), on average. In scenario

---

① A tolerable waiting time for a simple information query is about 2 s [156].
② For operating tasks, the waiting time should be within 15 s [157].

1.4, device 1 and 4 generated bigger messages, 264 (byte) on average. They hosted microservice $S_{C \to D}$ and $S_{B \to F}$, respectively, received discovery messages from the device who provided join-service $S_{DF \to Z}$. None of them can satisfy the goal of data D and F independently, and so device 1 and 4 cached information about the unfinished goal in their sending-out discovery messages. As a result, they sent bigger messages.

Table 8.3　Service provider's time consumption on each step in the GoCoMo service composition process, the average number of sent

Scenario	Device No.	Build time(ms)	Planning time(ms)	Execution time(ms)	Msg No.	Msg size (byte)
1.1	2	124	42	34	2	227
	3	63	40	31	2	227
	4	130	51	39	2	227
1.2	2	118	45	34	2	227
	3	58	43	36	2	227
	4	136	52	41	2	227
	5	80	50	29	2	227
1.3	2	121	41	30	2	227
	3	48	42	39	2	227
	4	131	47	36	2	227
	5	42	70	45	2	227
	6	130	40	72	2	227
1.4	1	52	24	47	2	264
	2	128	45	29	2	227
	3	46	55	32	2	227
	4	125	40	23	2	264
	5	45	38	49	2	227

Figure 8.4 illustrates the overall performance of GoCoMo in different scenarios, where Max. CPU usage represents the GoCoMo App's maximal CPU usage in composition processes, and Max. memory usage represents the maximal memory usage. The results show that GoCoMo's CPU usage was below 16.04%, and was even smaller on devices that had more computing resources. The most RAM-costly process occurred on device 6 (Motorola RAZR i) and 7 (SM-G920F). This is because that Motorola RAZR i has limited resources comparing to the rest of devices, and SM-G920F's processor (Galaxy S6) trades off CPU usage to RAM usage. Although they were less efficient, their memory usages were only a very small portion of the total, 1.1%. On the other hand, the GoCoMo App implemented GoCoMo's service provider and client modules in an integrated architecture. They have the potential to be implemented as adjustable

modules, by which a pure service provider does not have to load the modules that support a composition client's behaviour to reduces resource consumption on resource-constrained devices.

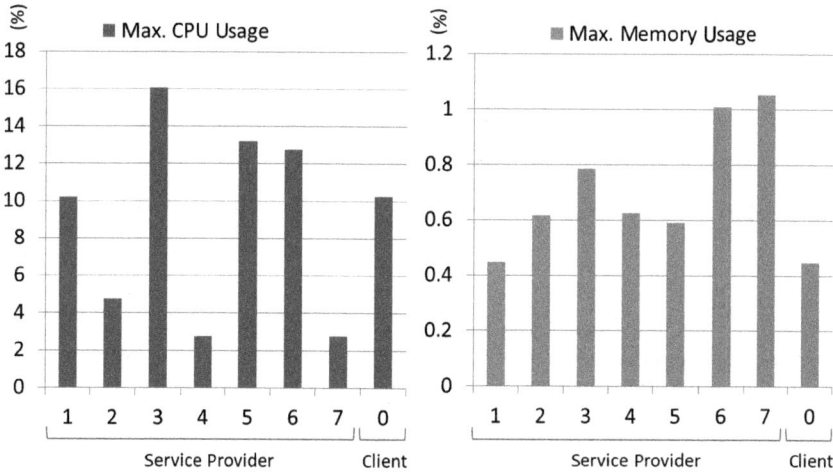

Figure 8.4   GoCoMo's maximal CPU usage and maximal memory usage on different devices

2. Adaptation Case

The adaptation case compared GoCoMo's adaptable composite against a service composition model using static composites, and measured composition response time and failure numbers. The results are shown in Figure 8.5. Overally, the failure rate for the two approaches increased with longer composites and more complex control logic. In addition, comparing to the results illustrated in Figure 8.3(a), GoCoMo produced more failures in scenarios 1.5-1.8 with dynamic service availability, but GoCoMo had less failures than the static composite approach. The static composite approach's response time in scenario 1.5-1.8 was similar to that of GoCoMo in scenario 1.1-1.4 as shown in Figure 8.3(b). However, GoCoMo, in scenario 1.5-1.8, took longer to return a composition result than the static composite approach. This is because GoCoMo adapts the execution path when a selected one becomes unavailable, which requires re-invoking microservices, taking extra time.

This case study demonstrated how GoCoMo addresses this book's challenges by supporting dynamic planning and self-organized service composition. The composition planning case showed that GoCoMo can reason about different service flows according to microservices available in the network, and the composition response time was acceptable. Given a slow messaging channel was used in this case, the response time has the potential to be further reduced by employing an advanced wireless protocol with fast transmission like WiFi or UWB[③].

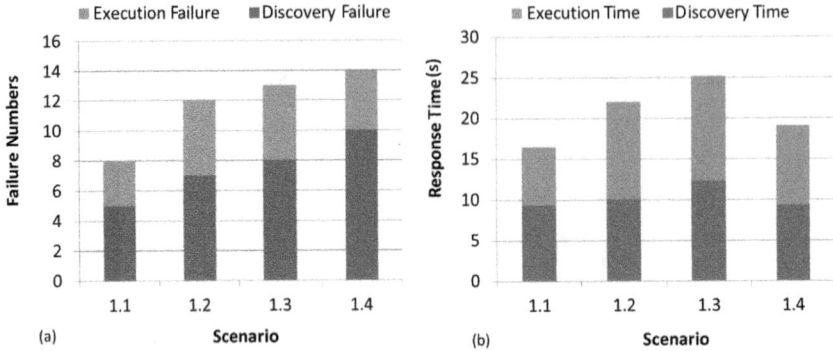

Figure 8.5   GoCoMo's feasibility on dynamic networks
(a) discovery and execution failures out of 50 attempts on a static network, (b) response time for the composition discovery process and the service execution process

As the case study adopted a small scale scenario, and a short distance transmission technology, this study did not include an analysis on GoCoMo's scalability or an investigation into GoCoMo's heuristic service discovery model. Moreover, the ad hoc network used in this study is static, so only unanticipated service availability was modelled, leaving the issue of dynamic network/service topology to be explored [122]. The global composition performance of GoCoMo in environments with dynamic network/service topology and the large-scale scenarios are analyzed using a simulation study, and will be presented in the next section.

8.3   Simulation Studies

Software-based simulations have been widely accepted in MANET research as an evaluation method [208]. Such simulations are low-cost, scalable and capable of modelling devices' different mobility behaviours.

8.3.1   Environment Configurations

This case study used ns-3 platform for its simulation, and deployed GoCoMo-ns3

---

③ A transmission on WiFi channel and on UWB channel can be about 75 times and 154 times faster than that on a Bluetooth channel, accordingly [217].

(See Section 6.6.2 and Appendix A.2) in a system with Intel Core i7-2600 3.4 GHz CPU, 8 GB RAM, running the Ubuntu 12.04.5 Desktop (32-bit).

1. General Settings

The experiment setting for this simulation were chosen according to recommendations[215] for MANETs. Specifically, the simulation used ad hoc on-demand Distance Vector Routing (AODV) for service routing and, for simplicity, considered only 2-dimensional 1000*1000 (m^2) terrain. The other settings like the number of nodes and communication distance refer to state of the art research for MANETs[23, 216], as shown in Table 8.4.

Table 8.4   Simulation configuration: general

General	
Simulator	ns-3
Clients	1
Communication range	250 m
Field	1000×1000 m^2
Microservice deployment	1 microservice per node
Semantic matchmaking delay	0.2 s [95]
Composition discovery	1-hop broadcast
Service routing	Dynamic AODV 10-hops
Sample	300 runs
Random	
Node placement	Random
Service execution time	0.01-0.1 s
Node movement	Random walk mobility model

Mobility Model and Node Topology

In the simulation, a random position was assigned to each node when an experiment gets initialized. The movement of a node during the experiment is controlled by a 2D random walk mobility model. In this mobility model, a node's movement is determined by a previously assigned speed as well as a constant time interval or a constant travel distance. At the end of the time or the distance, the network simulator calculates a new direction and speed for the node. All the nodes created in the simulation communicate using WiFi 802.11b in ad hoc mode. The communication range (distance) used in the simulation refers to outdoor WiFi communication distance, up to 250 m.

Microservices

The simulation created a number of nodes during initialization. One of them was a service client, and the rest were service providers. The simulation assumed a network with a low composition demand (one composition request each run), since the current version of GoCoMo does not provide a mechanism that addresses invocation failures caused by multiple composition processes competing for one service provider. Each service provider hosts one microservice, which is assigned a random execution time ranging from 0.01 s to 0.1 s. The number and the type of microservices are scenario-specific. Every individual evaluation scenario contains a number of different microservices and their duplicates, one per service provider (node). The simulation used alphabet transformation microservices that are similar to that were used in the case study introduced in Section 8.2.1. The simulation adopted a fixed semantic matchmaking delay [95] to minimize the variance performance of the GoCoMo composition process caused by different delays, easing the measurement of response time.

2. Evaluation Scenarios

GoCoMo was simulated with configurations that differ in their service density, node mobility and the complexity of service composition. These configurations included a set of controlled variables that define 4 different scenarios. Table 8.5 illustrates these scenarios marked from 2.1 to 2.4. The two columns on the left represent configuration parameters and their values.

Table 8.5   Scenarios for the simulation

Parameter	Configuration	Scenario			
		2.1	2.2	2.3	2.4
Environmental configuration					
Service density	20 (sparse)	★	★	★	★
	30 (medium-dense)	★	★	★	★
	40 (dense)	★	★	★	★
	50 (extra-dense)	★	★		★
Mobility (m/s)	0-2 (slow)	★	★	★	
	2-8 (medium-fast)	★	★	★	★
	8-13 (fast)	★	★	★	
Type of service flows	Sequential				★
	Parallel	★	★	★	★
	hybrid				★
Type of microservices	5				★
	10	★	★	★	
	15				★
Service instances per request	5 (simple request)	★	★		★
	10 (complex request)	★	★	★	
System configuration					
Service composition	Composition planning	★	★	★	★
	Execution		★	★	★
Heuristic level (interference degree)	5 (strong interference)			★	
	4			★	
	3			★	
	2			★	
	1 (weak interference)			★	
	0 (no interference)	★	★	★	★

Environmental configuration

The simulation defined service density ranging from 20 to 50 nodes and mobility including speed intervals 0-2 (human walking), 2-8 m/s (slow vehicles) and 8-13 ms (motor vehicles) m/s for random value taking. The type of service flows are the potential order and structure of discovered microservices that may be formed by a composition planning model according to these microservices' I/O dependency. Sequential service flows are the basic structure of a service composite, and contain only linear-services. Parallel and hybrid service flows include join-services, and are complex because of their control logic, but need to be flexible in some cases like when aggregation of data from different sources is needed. The simulation used different sets of microservices. For example, consider a scenario consisting of 5 different microservices, where a copy of each microservice can be deployed on one or more nodes in the network. In such a system, service composition could entail anything from 2-5 microservices involving multiple alternatives for each microservice.

The number of service instances per request refers to the complexity of a requested service composite, defined in composition requests. This simulation measured 5-service-instance composites and 10-service-instance composites in a scenario consisting of 10 different microservices. It also measured 5-service-instance composites in parallel execution flows in scenarios consisting of 5-15 different microservices. GoCoMo defines a composition request using input data and goals rather than conceptual composites (See **Definition 2** in Section 4.1) and assumes a client has no knowledge about the service availability in its local environment. So, generally a client cannot foresee the final composed service composite's complexity at the beginning (i.e., when issuing the request). The simulation controlled such complexity by initializing a service network containing a particular set of microservices, and measured network traffic and response time when resolving service composition in different complexities.

System configuration

To enable separate measurements on the performance of GoCoMo's composition planning model, execution model, and heuristic discovery model, the simulation also included configurations on GoCoMo itself. GoCoMo's heuristic discovery model (See Section 4.3) uses a value to represent the model's interference degree (See d in Equation 4.5), which indicates the model's influence on request flooding, and is determined by local network properties (e.g., the number of direct neighbours). In general, all the service providers (nodes) cannot have the same local network properties, e.g., equal in the number of neighbours, since the nodes are owned by third-parties, not distributed evenly, and dynamically change their positions. Service providers in the same network are likely to have different interference degrees, and the interference degree for each service provider changes over time. This simulation simplified this, and made all the service providers share the same interference degree to evaluate how varying interference degrees affect the GoCoMo composition model in different environments. This study used 6 degrees of interference signed from 0 to 5, and degree 0 means that GoCoMo has no interference on flooding, which means the request flooding is uncontrolled. Degree 5 represents the highest interference degree.

Scenario 2.1: Composition length, mobility and network density's influence on the flexibility of service planning

This scenario investigated how the composition length, mobility and network density impact the flexibility of service planning. This scenario used 10 different microservices, two levels of request complexity: 5-microservice and 10-microservice composite, and sequential service flows. GoCoMo is configured to support composition planning and using uncontrolled request flooding. In this scenario, planning failure rate, the number of sending messages and response time were measured.

Scenario 2.2: Composition length, mobility and network density's influence on the flexibility of service execution

This scenario explored how the composition length, mobility and network density impact the flexibility of service execution. This scenario had the same environmental configuration as Scenario 2.1, and configured GoCoMo to support a full functionality of service composition including composition planning and execution. Similar to Scenario 2.1, its request flooding is uncontrolled.

Scenario 2.3: Impact of heuristic service discovery

This scenario investigated how GoCoMo's different interference degrees affect the flexibility of the composition discovery process. This scenario used complex requests since complex requests lead to more service providers participating in the request resolving and, in turn, are likely to generate more discovery messages. Controlled flooding is designed to reduce message transmission, which in turn reduces message loss caused by packet collision[4]. This scenario employed all 6 interference degrees to demonstrate how the heuristic service discovery affects the failure rate and the system traffic in a set of environments.

---

[4] Packet collision will be introduced in following parts about adaptability of composite services.

Scenario 2.4: Impact of environment including complex service flows

Composites containing different service flows may increase the chance of more service providers being used to solve a request. It may also improve the quality of the final result This scenario explored the impact of environments including complex service flows.

8.3.2 Baseline Approach

To propose the establishment of a good baseline against which for GoCoMo to compare, this study combined state of the art functionalities. In particular, a decentralized cooperative discovery model [61] combined with a continuing message passing model [87] that enables decentralized service invocation. This baseline approach is referred to as CoopC in the following sections. Note that there are service composition approaches [21, 23, 30, 39] related to this research, but the simulation did not consider them as baseline approaches because their composition discovery processes are workflow-driven, which is inflexible (See Section 2.4.3).

CoopC's cooperative discovery employs a traditional backward goal driven service query and forward service construction mechanism. It generates a sequential service flow as the discovery result. However, CoopC's offline approach to service planning means that it does not support runtime service composite adaptation. Unlike the cooperative discovery model used as an input to CoopC [24, 61], this study implemented CoopC to start service execution when the first pre-execution plan is found. The plan is passed through the service execution path to indicate which microservice will be invoked for subsequent execution. This makes CoopC and GoCoMo more comparable when measuring response time. In addition, the latest version of CoopC assumes that a semantic service overlay network is previously cached to facilitate service discovery [24]. Given the infrastructure-less nature of our target network, this study went for an early version of CoopC [61] that uses request flooding for service discovery instead of the overlay infrastructure. CoopC's service execution was based on a decentralized message-passing model [87] that passes an invocation message from one microservice to its successive microservice. [87] recovers a failed execution process by giving the failed microservice a second attempt, which is controlled by a scope manager that maintains runtime status information about the microservices in its responsibility scope. CoopC implements this execution model [87] in an infrastructure-less network, so such a scope manager and the retry-based failure recovery was not adopted. Table 8.6 illustrates the difference and similarity between CoopC and GoCoMo.

Table 8.6   Comparison of the baseline CoopC with proposed GoCoMo

Items for Comparison	CoopC	GoCoMo
Composition planning	Traditional backward	Backward supports parallel
Service invocation	Decentralized	Decentralized
Request routing	Broadcast flooding	Heuristic broadcast
Dynamic binding	QoS-driven	On-demand
Fault tolerance	N/A	Dynamic recovery

8.3.3   Simulation Results and Analysis

1. Flexibility of Service Planning

This case study applied scenario 2.1, and measured planning failure rate in the simulated target environment, to quantify the flexibility of service planning. The failure rate and the GoCoMo composition planning model's performance are illustrated in Figure 8.6, Figure 8.7 and Figure 8.8. On finding pre-execution plans, GoCoMo shows

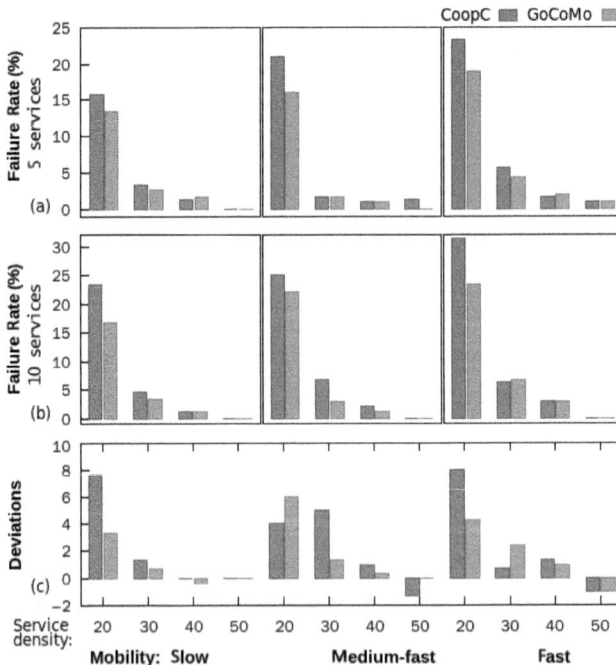

Figure 8.6   Planning failure rate in mobile networks
(a) 5 service instances per request, (b) 10 service instances per request, (c) the failure rate deviation between (a) and (b) [127]

a higher possibility of returning a pre-execution plan than CoopC, as shown in Figure 8.6(a) and (b). In particular, with the increasing complexity of service composition shows in Figure 8.6(c), GoCoMo raised about 0%-6% failures while CoopC raised approximately 0%-8% failures in most of the scenarios. The results also show that Go-CoMo discovery spent less time than CoopC to return the first pre-execution plan (Figure 8.7), but it sent slightly more messages than CoopC's discovery model to resolve a simple request in the sparse scenario (20 nodes) and the medium-dense scenario (30 nodes), as shown in Figure 8.8.

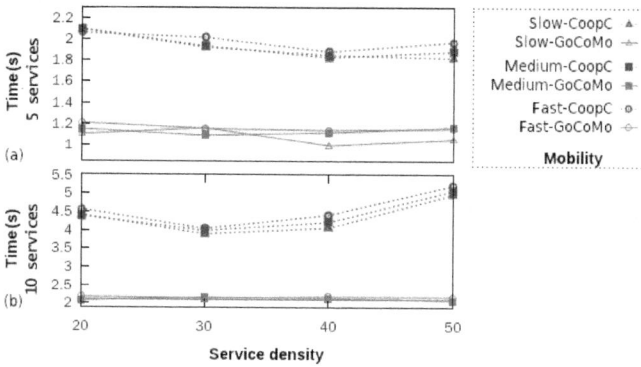

Figure 8.7   The discovery time in mobile networks
(a) 5 service instances per request, (b) 10 service instances per request [127]

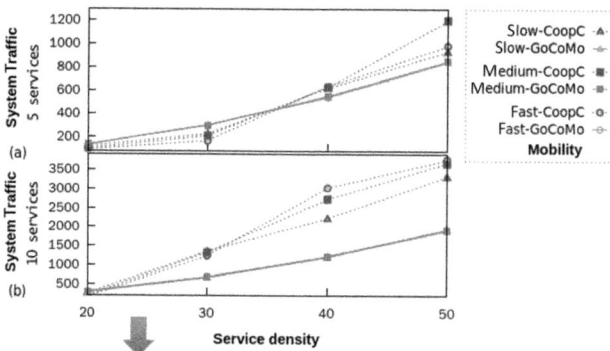

GoCoMo's system traffic:

	Slow Mobility		Medium-fast Mobility		Fast Mobility	
	5 services	10 services	5 services	10 services	5 services	10 services
20 providers	140	299	139	294	134	283
30 providers	324	701	323	701	321	704
40 providers	575	1268	586	1264	577	1272
50 providers	898	1996	902	1984	904	1993

Figure 8.8   The discovery traffic in mobile networks
(a) 5 service instances per request, (b) 10 service instances per request [127]

GoCoMo discovers more quickly, since it is not like its counterpart that requires one more step to finish the service discovery process, which constructs a pre-execution composite by forwarding a construction message to all the participant service providers after the backward service query. GoCoMo allows the fragments of execution plans to be selected and cooperate at execution time. GoCoMo produces slightly more traffic in some scenarios as it discovers more service links to find various possible execution paths for a single pre-execution plan.

2. Adaptability of Composite Services

A composite solution is adaptable if the system is able to compose solutions and complete service execution even in a mobile environment. The execution failure rate was calculated to show such adaptability for GoCoMo and CoopC in the case study. In this simulation, scenario 2.2 was used, and both of the approaches are implemented such that service execution starts immediately when the first pre-execution plan is returned to the client.

For execution failure rate (Figure 8.9) GoCoMo is more successful compared to CoopC in sparse networks (20 nodes) for most scenarios and also in dense networks (30-50 nodes). CoopC produced heavy system traffic during service execution (See Figure 8.10) when service density increases. This is because when the first returned plan is applied for execution, CoopC may still have participants that are performing service discovery (mainly in the forwarding process for service construction). Such system traffic occurs at any time less than t, for $t \in [0.55, 5.3]$ s, which indicates a frequent interaction between composite participants[5]. Frequent interactions in a network increase the possibility of high packet collision failures [67, 89]. Therefore, CoopC had more failures even though service density is increased. Although GoCoMo had the same tendency when service density increases from 40 to 50, in general, GoCoMo produced less failures than CoopC in dense networks. Starting service execution after the service discovery process leads to all participants reducing some interactions in the service execution process, which may prevent such frequent interactions, but it delays the service composition process, which may cause even more failures because of service path loss in a mobile environment [23, 79]. With a high service density (e.g., above 40 microservices), GoCoMo in a fast mobility network returned about 0% -17.33% more failures than in a medium-fast mobility network or a slow mobility network. In a low service density network (e.g., a network with 20 microservices), the failures increased to 4.67% -25.33%. This is because low service density networks only have a limited number of microservices in a node's communication range, which makes it hard for GoCoMo to find alternative service execution paths to replace failed paths.

---

[5] Service execution time interval [0.55, 5.3] s was calculated by removing the time spent on discovery (in Figure 8.7) from the response time (in Figure 8.11).

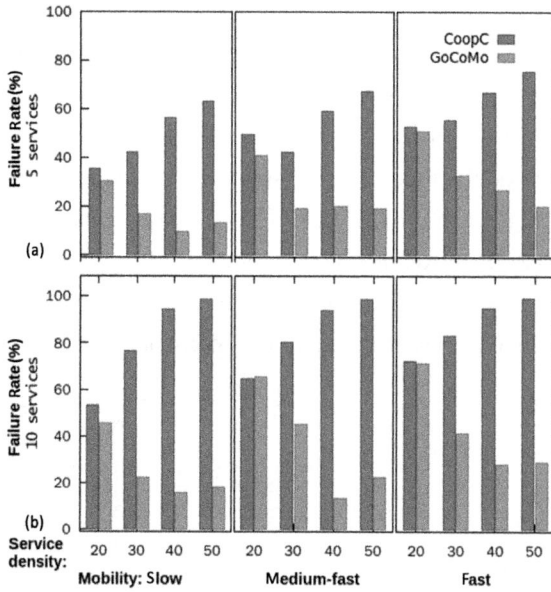

Figure 8.9   Execution failure rate in mobile networks
(a) 5 service instance per request, (b) 10 service instance per request [127]

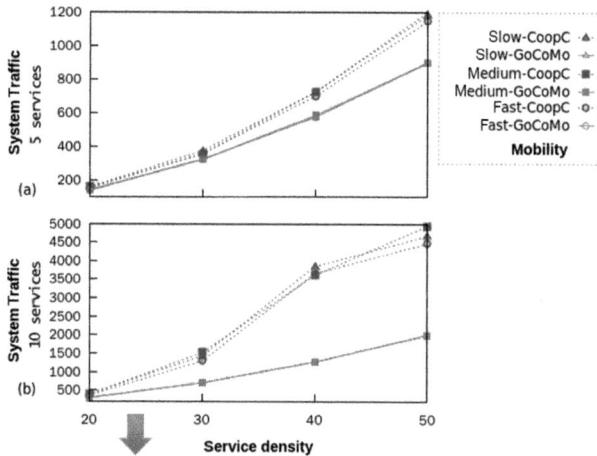

GoCoMo's system traffic:

	Slow Mobility		Medium-fast Mobility		Fast Mobility	
	5 services	10 services	5 services	10 services	5 services	10 services
20 providers	274	465	274	468	272	460
30 providers	497	808	523	802	520	808
40 providers	1112	1379	1137	1375	1121	1382
50 providers	1744	2213	1754	2201	1760	2224

Figure 8.10   The overall traffic in mobile networks
(a) 5 service instance per request, (b) 10 service instance per request [127]

For service execution performance, the system traffic for a composition process was counted and the response time was measured. The results (Figure 8.10 and Figure 8.11) show that GoCoMo processes service composition more quickly than CoopC approach in high density scenarios and is less affected by service density. In highly dynamic (fast) networks, the time spent on service composition for GoCoMo increases slightly faster than that in slow networks. This is because execution failure recovery requires jumping back to an executed node. Figure 8.10 shows that GoCoMo generates less traffic compared to CoopC, because CoopC merges all the partial plans during service discovery, which increases interactions among participants.

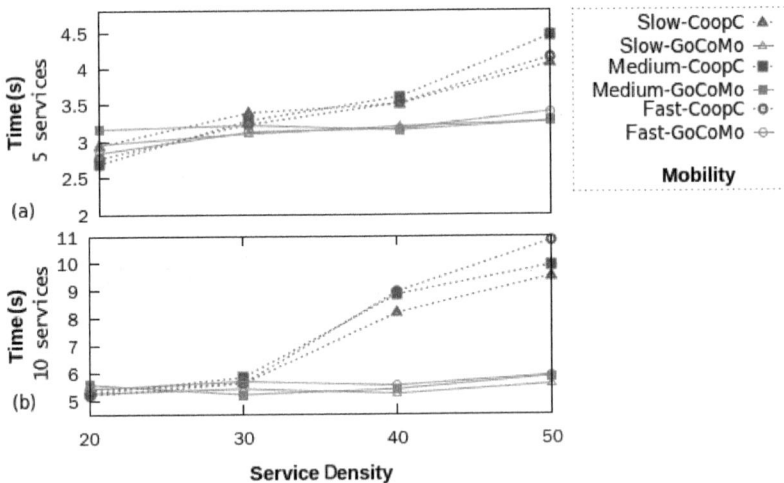

Figure 8.11   The response time in mobile networks
(a) 5 service instance per request, (b) 10 service instance per request [127]

3. Impact of Heuristic Service Discovery

GoCoMo applied a heuristic service discovery mechanism. The selection of the interference degree may affect the results for returning a composition solution. A low interference degree can lead to more discovery messages that may increase the possibility of packet collisions and packet loss [89], and in turn composition failures, while a high interference degree in some cases may result in a limited discovery scope, and are likely to increase discovery failures. A test used scenario 2.3 was run to measure this influence, which assumed all the nodes in the network use the same interference degree for simplicity. The test applied 6 levels of heuristic discovery from 0 to 5. Level 0 means there is no interference, and level 5 process uses the highest interference degree. Figure 8.12 shows that, in a medium-dense network (30 microservices) with medium-fast mobility, a high discovery interference (i.e., 5) will reduce composition failures. This is because, in such a dense network, even a small search scope can support enough backup execution paths for failure recovery, while it sends less query messages than its high-level counterparts, which reduces the potential for packet collision failures.

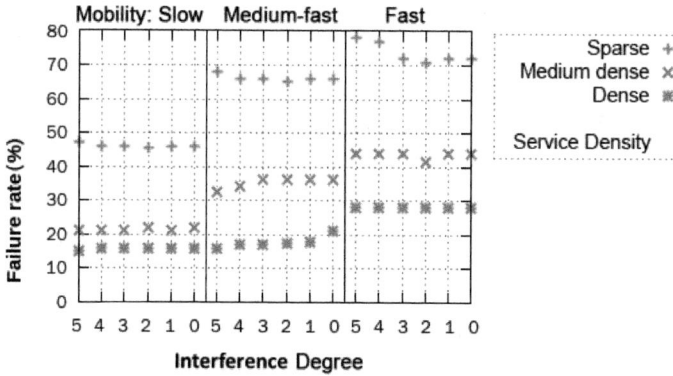

Figure 8.12   Interference degrees affect failure rates [127]

Figure 8.13 illustrates the system traffic in a medium-fast network with different interference degrees and service densities. As shown in Figure 8.8 and Figure 8.10, service providers' mobility has very limited impact on GoCoMo's system traffic, and so this simulation only presented GoCoMo's system traffic in a medium-fast network as a representative case. The result shows that a high discovery interference can reduce system traffic, especially in a dense service network. However, in a sparse service network, a high interference degree can increase composition failures (See Figure 8.12), as it may imply a comparatively small search scope.

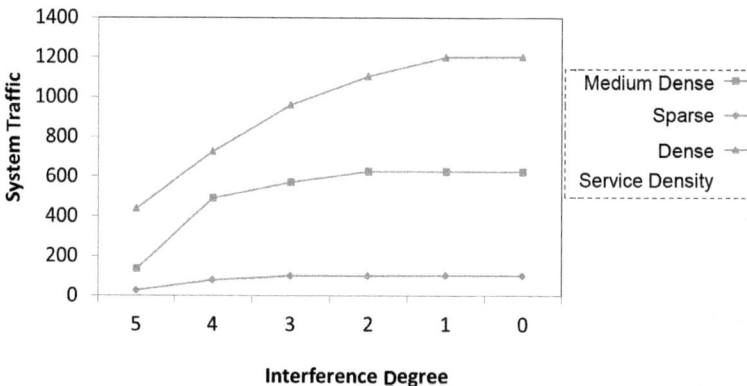

Figure 8.13   Interference degrees affect the system traffic [127]

4. Planning Complex Service Flows

The evaluation scenario (scenario 2.4) for assessing the support for complex service flows contains one client node, 5-15 different microservices and their duplicates, one per service provider. These microservices vary in type and number of their input and output parameters. This reflects that when service instances engage in a workflow, each may have multiple, different in-degrees and out-degrees[⑥]. The potential service flows constructed by these microservices for a composite service are illustrated in Figure 8.14. These service flows connect participating microservices relying on their data dependency, and include sequential workflow models (model a), hybrid workflow (model b) [158, 218], and parallel workflow models (model c and d). Each of them includes 5 service instances. All the service flows start and end with the client node that issues a service query, looking for a service composite that converts data x to y. GoCoMo's failure rate on scenarios that contain the different kinds of workflow models is shown in Figure 8.15. It was simulated under medium-fast networks. On returning solutions (See Figure 8.14), GoCoMo supported scenario c the best, and had more failures in the scenario with type-d microservices. CoopC supported only type-a (sequential) microservices, and so only GoCoMo results are illustrated in Figure 8.15.

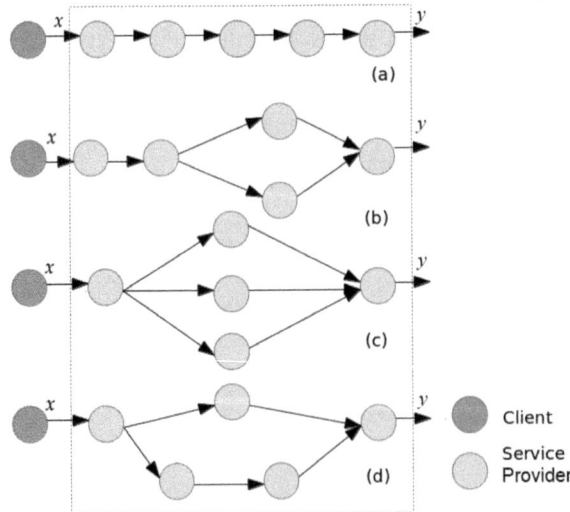

Figure 8.14 Potential service flow for a composite service that supports data transition: $x \rightarrow y$ [127]

---

⑥ In-degree and out-degree indicate the number of connections entering and leaving a workflow node, respectively [218].

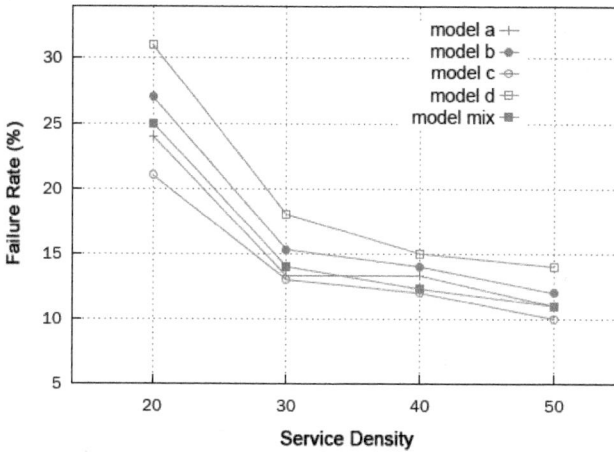

Figure 8.15   Failure rate for the data transition ($x \rightarrow y$) request [127]

5. Availability of Cooperative Microservices Provisioning

In this section, we investigate the performance of the algorithm LOCASS. In LOCASS, we set social density $\eta = 0, 0.3, 0.6, 0.9, 1$ respectively. Specifically, LO-CASS with social density $\eta = 1$ can be considered as the full cooperation case where each device altruistically devote its cache space to maximize the system performance, specifically the average offloading ratio. LOCASS with social density $\eta = 0$ can be considered as the full selfishness case where each device just stores contents only for itself and has no right to access others' cache space. We compared the performance of LOCASS with other three caching algorithms: Most Popular Content (MPC) caching, Random Caching and Greedy caching. In MPC caching, each device just stores the most popular contents according to its individual request probability. In Random caching, each device randomly chooses contents to store until its own cache space is full. The Greedy caching starts with an empty set, at each step, it adds one content with the highest marginal average offloading ratio to the set until its cache is full. In this part, only the performance of those three caching algorithms when social density $\eta = 0.3$ is shown.

We adopt two types of social graph models for depicting the social relationship among devices:

- Erdos-Renyi (**ER**) graph model [219]: Each pair of devices has a fixed probability of being cooperative neighbours or not. All devices are likely to have similar degrees.

- Scale-free (**SC**) graph model [220]: The degree of devices follows power law distribution where only a few of devices are equipped with high degrees and the rest of them are equipped with low degrees.

Then, we generated the number of contacts within unit time λ for each pair of devices according to a Gamma distribution Γ (4.43, 1/1088).

Additionally, we assumed that the content request probability follows the Zipf distribution with parameter α, i.e., $P_{m,f} = \dfrac{(f)^{-\alpha}}{\sum_{f' \in \mathcal{F}} (f)^{-\alpha}}, \forall m, f$. Other baseline simulation parameters were shown in Table 8.7.

Table 8.7　Baseline simulation parameters

Parameters	Quantilies
Total number of devices M	200
Total number of contents F	200
Cache size of each device S_m	50 MB
Parameter of Zipf distribution α	0.6
Parameter of **ER** social graph model η	0.3
Transmitted amount of data during one contact	20 MB
Deadline time T_D	200 s
Size of each content r_f	100 MB
Social tie strength ω	1

Virtual cache space is a set of individual cache space which is associated with the device number and the individual cache size. Figure 8.16 and Figure 8.17 shows the performance of the caching algorithms by increasing the individual cache size and the device number, respectively. As the virtual cache space increases, the average offloading ratio of all the three algorithms increases except for the MPC with social density $\eta = 0.3$ and LOCASS with social density $\eta = 0$. The increasing of average offloading ratio is because more content copies can be stored and shared via D2D links as the virtual cache space increases. In the MPC with social density $\eta = 0.3$, each device stores the same content so that no performance improvement can be achieved with the increasing device number. In LOCASS with social density $\eta = 0$, each device is fully selfish and cannot access other devices' cache space, so the average offloading ratio doesn't change with increasing device number. In addition, in terms of average offloading ratio, LOCASS always outperforms MPC caching, Random caching and Greedy caching no matter how the device number or cache size changes. Greedy caching outperforms the MPC and Random caching when the virtual cache space is small. And Random caching outperforms MPC caching when the virtual cache space is large.

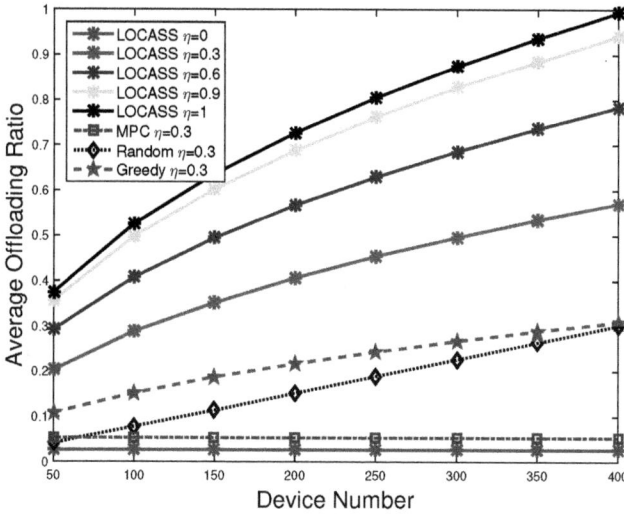

Figure 8.16   Average offloading ratio as the virtual cache size space changes in ER model-Average offloading ratio v.s Device number [171]

With the increasing social density η, each device can access more cache space, and its requests can be taken into consideration by more cooperative neighbours. So, the average offloading ratio of LOCASS improves with larger social density η. From the Figure 8.16, with the increase of social density η, LOCASS utilizes the virtual cache space in a less effective way in that the same size of accessible virtual cache space can achieve a lower average offloading ratio. For example, when the cache size is 50 MB and the device number is 200, each device of LOCASS with social density $\eta = 0.3$ can access 30% of virtual cache space and the average offloading ratio is 40.64%. Each device of LOCASS with $\eta = 1$ can access 100% of the virtual space, and the average offloading ratio is 72.73%. In Figure 8.17, we can also find that the gap between different cases of LOCASS becomes larger with the increase of virtual cache space before the average offloading ratio reaches to 1.

In the Zipf distribution $P_{m,f} = \dfrac{(f)^{-\alpha}}{\sum_{f' \in \mathcal{F}} (f)^{-\alpha}}, \forall m, f$, a larger α means a steeper

request probability distribution where the majority of requests probably concentrate on the limited number of popular contents.

Figure 8.18 depicts the simulation results of the average offloading ratio of the three algorithms with the change of the Zipf parameter α. With the increase of the Zipf parameter α, the average offloading ratio of three algorithms increases except for Random caching. In addition, LOCASS outperforms MPC caching, Random caching and Greedy caching no matter how α changes. Greedy caching outperforms Random caching when α is large. And Random caching outperforms MPC caching when α is small.

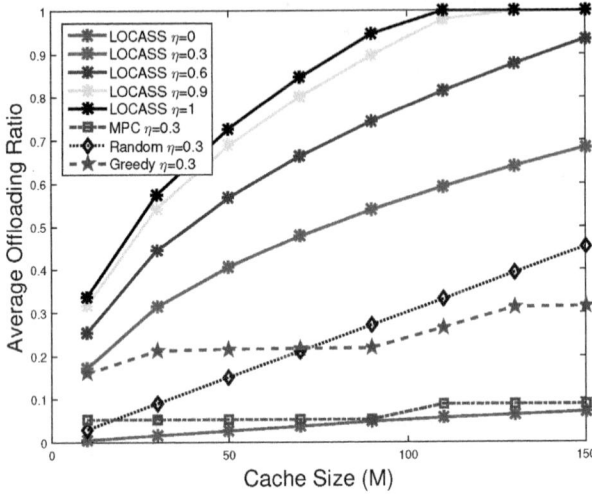

Figure 8.17 **Average offloading ratio as the virtual cache size space changes in ER model-Average offloading ratio v.s Cache size** [171]

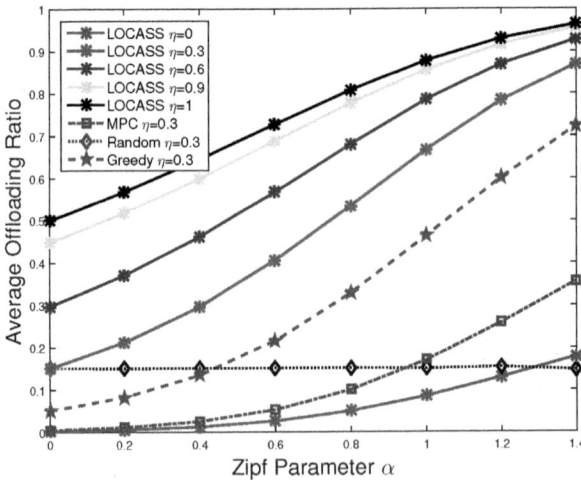

Figure 8.18 **Average offloading ratio as the Zipf distribution parameter α changes** [171]

Considering the rising speed with the increase of the Zipf parameter α, the average offloading ratio of LOCASS rises slower with the increase of social density η. Because devices can access limited virtual cache space, only a few numbers of popular contents can be stored (See Figure 8.16). With the increase of social density η, the number of the popular contents increases. The sum of the probability of those popular contents being requested increases slowly, thus smaller performance improvement can be achieved when α increases. Therefore, LOCASS is more sensitive to the change of the content request probability distribution with the decrease of social density η.

Figure 8.19 depicts the effect of social tie strength on the average offloading ratio. With the increase of social tie strength, the average offloading ratio of LOCASS improves except when social density $\eta = 0$, 1. The increasing of average offloading ratio is because the cooperation between devices is enhanced, and devices are more likely to store contents for cooperative neighbours. As the social tie strength ω increases, those contents with comparatively lower popularity have large chance to be cached and increase devices' social selfishness-based utilities. In the LOCASS with social density $\eta = 0$, each device cannot access cached contents from other devices so that social tie strength ω has no impact on the average offloading ratio. Besides, in terms of average offloading ratio, LOCASS outperforms MPC caching, Random caching and Greedy caching no matter how social tie strength ω changes when social density $\eta = 0.3$. We can also find that the average offloading ratio of LOCASS converges faster with the increase of social density η. In Figure 8.19, the average offloading ratio of LOCASS with social density $\eta = 0.3$, 0.6, 0.9, 1 converges when ω goes beyond 0.5, 0.2, 0.2 and 0.1 respectively. When the average offloading ratio converges, it also means that the cached content distribution will not change. Therefore, when the social tie strength ω is large enough, a large scale of rearrangement of content won't happen if the social tie strength changes a little. Larger η means more cooperative neighbours and larger cooperative utility for each device. Smaller social tie strength ω can pose the same effect on the average offloading with the increase of social density η.

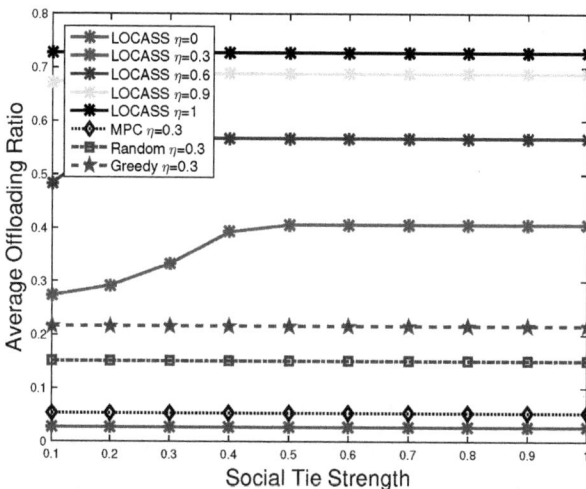

Figure 8.19   Average offloading ratio as the social tie strength ω changes in ER model [171]

Therefore, the increase of social tie strength ω can enhance the cooperation of devices and improve the average offloading ratio in LOCASS. With more cooperative neighbours, i.e., larger social density η, social tie strength ω poses less impact on the average offloading ratio.

Larger social density η means more cooperative neighbours for each device. In another word, more virtual cache space can be accessed and more devices will take the requests of device m into consideration. We investigated the evolution result with the increasing of social density η in Figure 8.20 and Figure 8.21, the average offloading ratio of LOCASS outperforms MPC caching , Random caching and Greedy caching no matter how social density η changes. In addition, the average offloading ratio of the four caching algorithms improves with the increase of social density η except MPC caching. For LOCASS, the increasing speed reduces due to the fact that more cache space is utilized to store those comparatively unpopular contents, which is shown in Figure 8.17. Small performance improvement can be achieved if those comparatively unpopular contents are stored. Thus, the increasing speed of the average offloading ratio decreases as social density η grows.

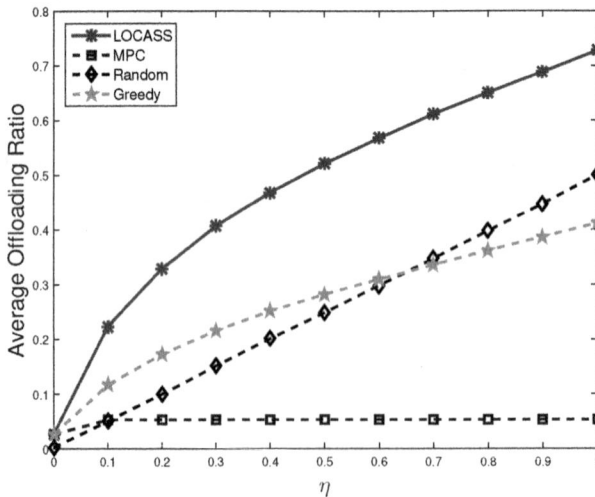

Figure 8.20   The performance as the social density η changes in ER model (Average offloading ratio as the social density η changes [171])

Figure 8.21 depicts the Cumulative Distribution Function (CDF) of the virtual cache space volume for each content with different ranks in LOCASS. The content rank is sorted by content request probability in descending order. The content with a smaller rank is equipped with a larger request probability. If some new contents are requested, whether a large scale of content rearrangement will happen depends on the rank of those the new contents. If the new contents are with small ranks, a large scale of virtual cache space will be utilized to store, and a large scale of rearrangement will happen. Correspondingly, if the ranks of new contents are large than 100, those contents have no chance to be stored so that the rearrangement of content won't happen. As shown in Figure 8.21, the whole cache space is mainly occupied with the small-rank contents. Besides, with the increasing social density η, each individual cache space can be accessed by more devices, so fewer of copies can achieve similar performance. Devices with a larger η are likely to store the popular contents as well but tend to store those comparatively unpopular contents. This can further improve the system performance. Therefore, a larger social density η means more cache space of sharing via D2D links, and devices are encouraged to store diverse contents to avoid duplicate caching.

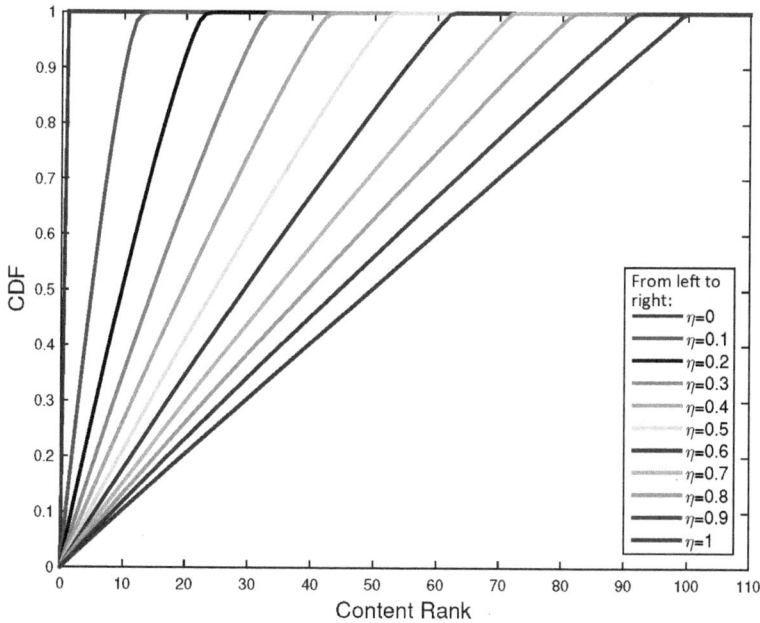

Figure 8.21   The performance as the social density η changes in ER model (Cached content distribution as the social density η changes for the LOCASS [171])

To further validate the performance of LOCASS, we utilized the real trace of the Infocom06 dataset [221] to conduct a trace-driven simulation. The Infocom06 dataset contains the contact logs among 79 candidates in a period of 337 417 seconds [221]. In LOCASS, we applied two social graph model to construct social ties among these candidates: ER model and SC model. We assumed that the sum of social ties for the two models were equal. We set social density $\eta = 0.3$. For each tie, the social tie strength ω was generated randomly in the interval [0,1]. Here, we just show the cached content distribution of devices with different degrees.

Figure 8.22 shows the evolution result for the four cases (i.e., LOCASS $\eta = 0$, 0.3, 1 with ER model and LOCASS $\eta = 0.3$ with SC model) with the change of deadline time T_D. Except for the case LOCASS $\eta = 0$ with ER model, the average offloading ratios of the other three cases grow along with the increase of deadline time T_D. That is because a larger T_D means more contacts between each pair of devices, and more cached segments can be transmitted via D2D links. In the case of LOCASS $\eta = 0$ with ER model, devices cannot access any other's cache space, so the average offloading ratio doesn't change with the increase of T_D. If there is a same social density η, when T_D is small, the average offloading ratio of SC outperforms that of ER. Since in the ER model, all devices' degrees are low, and the cooperative utilities of devices with small T_D are comparatively small. With a small cooperative utility, there is a lack of cooperation with their cooperative neighbours and poor performance can be posed. However, in SC model, there are a few devices equipped with high degrees, and their cooperative utility is comparatively large. With the large cooperative utility, those devices with high degrees have a strong motivation to cooperate with their cooperative neighbours. Therefore, the average offloading ratio of SC outperforms that of ER when T_D is small. In Figure 8.22, when T_D exceeds 800 s, for each pair of cooperative neighbours, almost all of the cached contents can be transmitted before reaching the deadline time. In such situation, a large part of the virtual cache space in LOCASS with SC model is utilized to store unpopular contents for cooperative neighbours with high degrees, which achieve only a limited performance improvement. However, in ER model, all devices have similar degrees and tend to store popular contents, which will achieve a great performance improvement. Therefore, in terms of average offloading ratio, LOCASS with ER model outperforms LOCASS with SC model when the deadline time T_D is large.

Figure 8.23 shows how cache spaces are occupied by different contents, represented by a CDF graph. From the system perspective, LOCASS-SC utilizes more cache space to store contents with a small rank than LOCASS-ER. This is because of the devices in SC model equipped with low degrees, in that only a small part of virtual space can be accessed by each of them. Devices with a low degree can access few segments from their cooperative neighbours and have to store popular contents for themselves. Therefore, compared to LOCASS-ER, LOCASS-SC utilizes a larger part of cache space to store popular contents. Considering the diversity of degrees in SC model, we also show the cache space occupied by different contents of our solution in the situation of devices with high degrees and low degrees which is called as high degree-SC and low degree-SC respectively. The former selected five devices with highest degrees, and the latter evaluated five devices with lowest degrees in SC model.

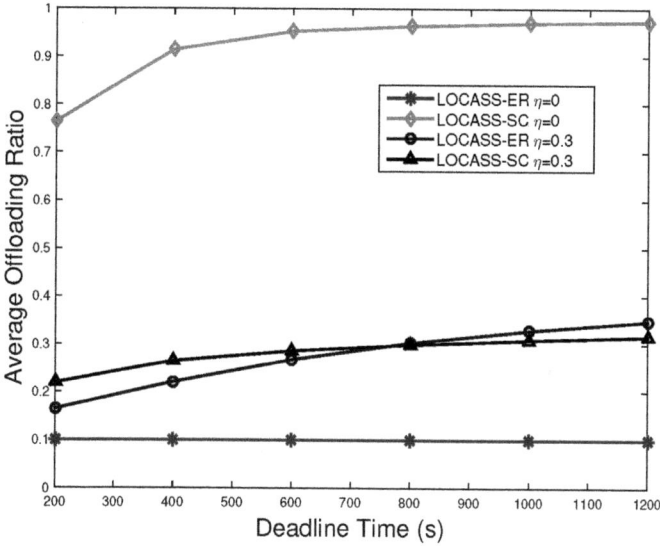

Figure 8.22　Average offloading ratio as the deadline time T_D changes in Infocom06 dataset [171]

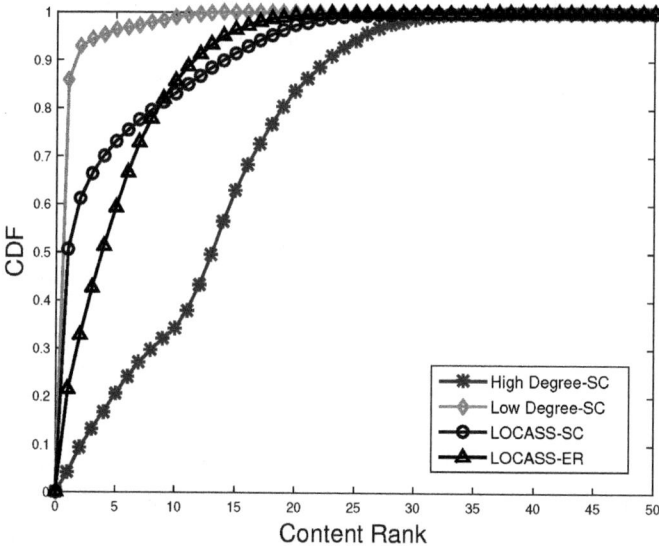

Figure 8.23　The cached content distribution in Infocom06 dataset [202]

As shown in Figure 8.23, devices with low degrees in SC tend to store popular contents, a.k.a., the contents with a small rank. In addition, devices with high degrees in LOCASS with SC tend to store comparatively unpopular contents and access popular contents from their cooperative neighbours.

8.4 Evaluation Summary

8.4.1 Service Composition

This chapter outlines GoCoMo's evaluation, including a prototype case study and a simulation-based experiment. The evaluation used a set of criteria driven by this book's challenges. In the prototype case study, GoCoMo's feasibility in dynamic environments that have no conceptual composite or composition infrastructure was demonstrated. The simulation demonstrated GoCoMo's success in the target environment, indicated by planning and execution failure rate, response time and system traffic.

The case study in this chapter measured CPU consumption and memory usage, and took them as the main factors that affect a model's feasibility on mobile devices. It would also be interesting to measure service providers' energy consumption as the battery is a constrained resource to mobile devices. However, the case study used an Android device monitor running on a desktop, which requires a device to be plugged into the desktop's USB port. This means all the devices were charged during the experiment, and so battery consumption was not measured. On the other hand, the major energy consumer was brought by wireless communications and screens. The GoCoMo App used in the case study keeps the screen active to display runtime information for evaluation, but it can be implemented as a background task. The power consumed when sending a message over a Bluetooth channel is about 0.145 J/Kb [217]. The case study results show that the required message communication is 2 messages $\times$ 227 (byte) on average, and so the GoCoMo composition process only requires a very small battery power.

In the simulation result, GoCoMo showed a reduction in communication over a selected related approach, and with an average of 2 messages for each participant using controlled request flooding. In the worst case (i.e., uncontrolled request flooding in Figure 8.13), there was an average of 10 messages per node in a sparse network, 20 messages in a medium dense network, and 32 messages in a dense network sent during a service composition process. The power consumed when sending a message over a WiFi network channel is about 0.012 J/Kb [217]. This chapter concludes that reasonably small battery power is required for the GoCoMo composition process.

Scenarios with a high service composition demand were not investigated. Such scenarios indicate multiple clients. GoCoMo uses on-demand binding and does not check a microservice's availability immediately before invoking the microservice. Service invocation is likely to fail when the required microservice is currently occupied by another composition process. As a service composition process locks a service provider's resource only when it is executing the required microservice, the length of a microservice's unavailability depends on the duration of the microservice's execution, which is normally predictable. Depending on a prediction of unavailable time, queuing up invocation messages when the time is short or otherwise re-selecting a service provider may solve the service invocation failure led by occupied service providers.

8.4.2 Cooperative Service Provisioning

Simulation results are provided to demonstrate the validity of LOCASS. The results show that LOCASS can offer much better performance, in terms of average offloading ratio, than traditional Random, MPC and Greedy caching algorithms. As the social tie strength and social density increase, the cooperation among cooperative neighbours is enhanced, and the average offloading ratio of LOCASS also improves. Additionally, in LOCASS, devices with low degrees are likely to store popular contents, and devices with high degrees tend to store those comparatively unpopular contents and fetch popular content from their cooperative neighbours.

Chapter 9

Discussions and Conclusions

THE research presented in this book has investigated a novel model for software service provisioning in mobile and pervasive computing environments. In particular, the research has focused on providing solutions to the challenges of composing services from multiple mobile devices, which behave in a decentralized fashion and adapt to network changes. A novel service composition model named GoCoMo that targets several service provisioning issues occurring in open and dynamic pervasive environments was proposed.

This chapter summaries the achievements of the research and its contributions, discusses GoCoMo's trade-offs, and highlights potential areas for future work.

Introduction

Chapter 1 describes the motivation of this book that arose from new challenges of service composition in pervasive computing environments. This book is especially concerned with the openness and dynamism features of the environments. The chapter argues that requests for complex functionality in a smart public space can be supported by the composition of services from a number of mobile devices located in the vicinity of the requester. However, such a dynamic, open computing environment can negatively impact the service composition process. Challenges include inadequate conceptual composites, limited system knowledge, unpredictable services availability, unreliable wireless communications, and dynamic composition links, making the discovery scope limited or service execution unstable, and therefore, composition failures can occur. The following chapters present research and case study results to meet these challenges.

Design Concepts

Chapter 2 analyzes the most related solutions and claims that the existing service composition model are not flexible enough to fully address these challenges. Based on the analysis on the state of the art solutions, a hypothesis is proposed, which assumes that a novel service composition model could reduce the composition failure probability, by dynamic composition planning that engages a big scope of potential service providers, through heuristic services discovery that reduces unnecessary system traffic, and using adaptable workflow-based composites that enable flexible execution and fast but low-cost failure recovery. The chapter reviews state of the art research in depth, and uses a taxonomy that covered the most related aspects of a service composition process, containing when and how to locate a provider, what routes a request to a destined service provider or a composition manager, how to resolve a composition request, when a service will be bound for execution, how to execute services in decentralized ways, and when and how a failure can be recovered. As a result, the chapter claims that the combination of dynamically controlled request flooding, goal-oriented planning, on-demand service binding, self-organized composition and dynamic failure recovery can provide a promising solution to the research presented in this book. However, existing approaches fail to bridge the gap between the solution and the target environment, as either they are inflexible about service discovery and planning or they have limited adaptability.

Microservice Deployment

Chapter 3 introduces the edge computing environment and the challenges when deploying microservices in it. The nature of such an environment includes latency-sensitive, openness, mobility, dynamism, and constrained power. Fog as a service model and the supporting technologies are presented to enable flexible service provisioning in edge networks. This chapter also investigates the adaptability requirement and solutions at the network edge. Representative pervasive applications on the current edge computing environment are introduced. This chapter claims that the distributed adaptation, scalability, reconfigurability, and the intelligent support are still unsolved problem in this domain.

Microservice Composition Model

Chapter 4 presents GoCoMo. To address the above issues, chapter 2 outlines a number of design requirements that GoCoMo must fulfill, and Chapter 4 introduces how to design GoCoMo using a set of design concepts that entirely satisfy the requirements. More specifically, first, GoCoMo designs a planning-based composition announcement that allows service providers to update and pass a composition request to discover a service composite hop by hop, and only announces the finished composite to the requester. Second, GoCoMo dynamically controls request flooding according to the density of each service provider's neighbour services and the global service discovery and execution's time constraints, which trades off discovery scope with system traffic. Third, a flexible backward planning is designed to enable parallel and hybrid service flows when they are necessary. Fourth, GoCoMo selects service providers based on estimated path reliability. Such path reliability values are calculated during service discovery and differ in every participant. Fifth, GoCoMo uses dynamic service binding and only locks a service provider's resource during the service's execution. Sixth, to adapt a composed service and engage the newly emerged service provider, GoCoMo allows service providers to proactively announce their available services. Composite participants receive such announcements and invite appropriate service providers to participate in a composition process. At last, service providers in GoCoMo only maintain execution paths linking to their subsequent services. The admission of service execution control is passed from one service provider to another.

Cooperative Microservice Provisioning

Chapter 5 presents a cooperative microservice provisioning at end devices, which takes the social selfishness into consideration. Mobile devices work together to solve user requests. They can be provided by different third parties according to the fog as a service model. Therefore, it is hard to define if they behave with more cooperation or selfishness. Social selfishness is introduced as a conditional cooperation behaviour. To simplify the question, this chapter uses caching service as an example and introduces a game.

Implementation

The implementation of the proposed technology is presented in Chapter 6 and Chapter 7. GoCoMo is realized as an application layer middleware that connects services that are deployed in different devices by composing them according to their I/O dependency. The middleware's modules and the inner-module implementation are specified in these chapters, which also includes two prototypes of GoCoMo, GoCoMo App and GoCoMo-ns3, that implemented the modules in the GoCoMo middleware, and were used for evaluation.

Evaluation

The evaluation of GoCoMo is described in Chapter 8, which shows GoCoMo's suitability and limitations in dynamic pervasive computing environments, and includes two detailed evaluations, a prototype case study and a simulation. The prototype case study deployed GoCoMo App in a testbed composed by a set of real mobile devices, measured its performance on each individual devices, and demonstrated that GoCoMo can flexibly compose a number of diverse functionalities (services) deployed in diverse Android-based mobile devices to support a user's request. The simulation adopted 4 different scenarios and ran GoCoMo-ns3 in these scenarios. Results illustrated that GoCoMo is more flexible in terms of composition discovery/planning, and service execution, comparing to a service composition approach that also supports composition planning (baseline). In addition, GoCoMo produced less system traffic and spent less time on the service composition process in the evaluation. However, GoCoMo, similar to the baseline approach, does not solve the invocation collision problem that may occur when the demand for composition is high.

In summary, this book has presented a novel service composition model that does not rely on a conceptual composite, uses only local service knowledge, has low composition failure probabilities, and produces acceptable system traffic in environments where service providers are mobile, and service availability are unpredictable.

Appendix A

Further Implementation Detail: Prototypes

A.1　GoCoMo App

GoCoMo App is an Android application that supports service provisioning in bluetooth-based ad hoc networks. It includes a prototype implementation of the Go-CoMo middleware and a visualization module, making service provisioning light-weight and flexible. GoCoMo App supports two groups of users: composition clients and service providers. The main features of GoCoMo App are illustrated below:

- Supports most of GoCoMo's functionalities and behaviours, including compo-sition request generation and issuing for composition clients, local composition planning, decentralized composition execution and adaptation, and service ad-vertising/inviting.
- Compatible with Android devices with API level 16 and above.
- Visualizes message transmission and the local composition process.

Figure A.1 illustrates the class diagram of GoCoMo App. The message builders are presented in Figure 6.8. The biggest rectangular in Figure A.1 wraps the group of classes introduced in the GoCoMo middleware, and the rest of the classes and packages support functions like wireless communication and process visualization.

GoCoMo App did not implement GoCoMo's heuristic discovery model because BlueHoc, the ad hoc network used in GoCoMo App, does not support multi-hop trans-mission or dynamic routing. GoCoMo App can be extended to support the heuristic discovery model in Android 5.0 Lollipop (API 21) and above versions as these An-droid versions allow applications to directly control Bluetooth for broadcasting. Go-CoMo App used JSON data to store the **Guidepost** object, and XML-based descrip-tions to specify GoCoMo messages. Figure A.2 and Figure A.3 illustrate an example of a running service provider and a running composition client, respectively.

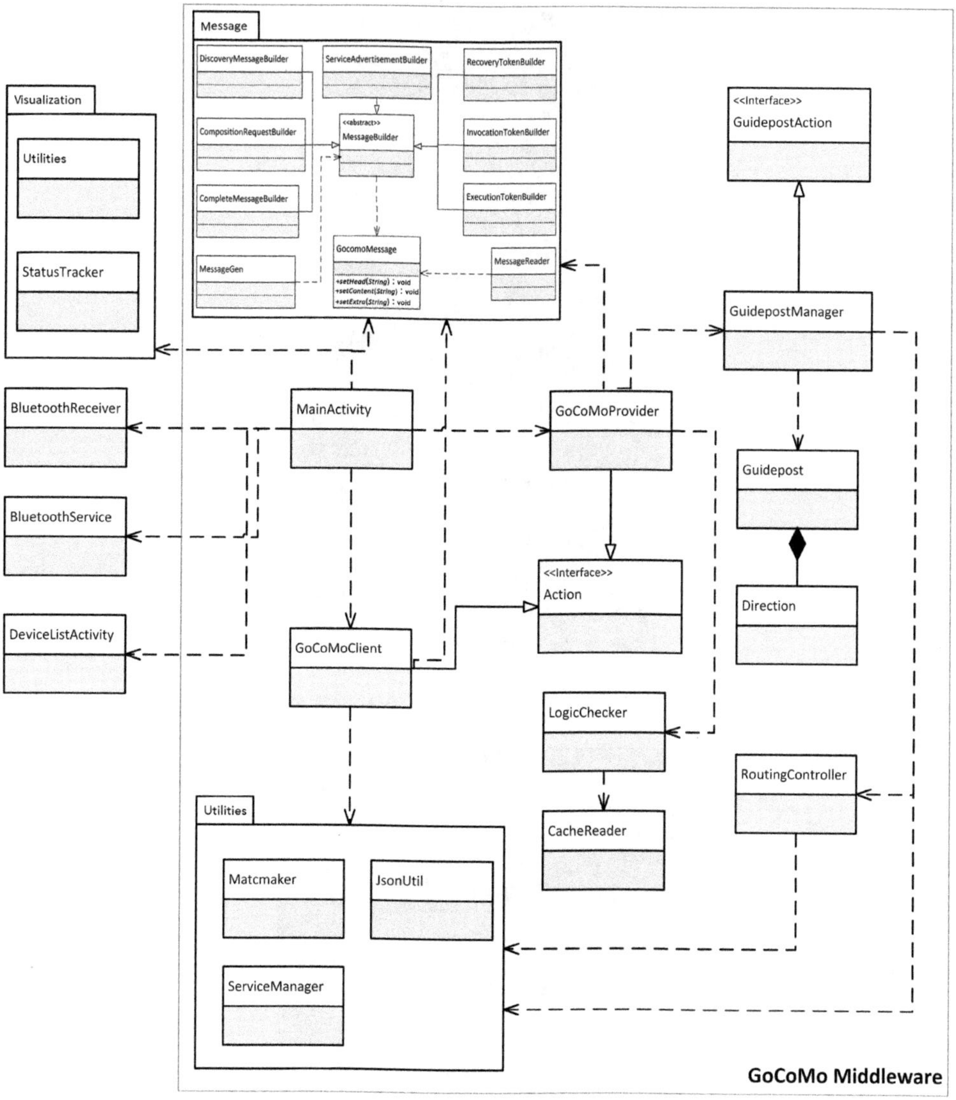

Figure A.1   GoCoMo App's class diagram

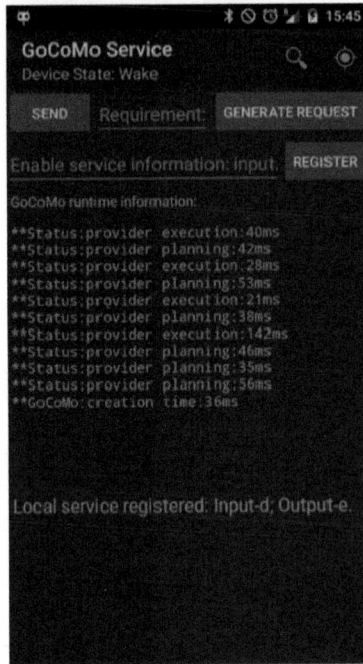

Figure A.2　GoCoMo App: a running device that acts as a service provider

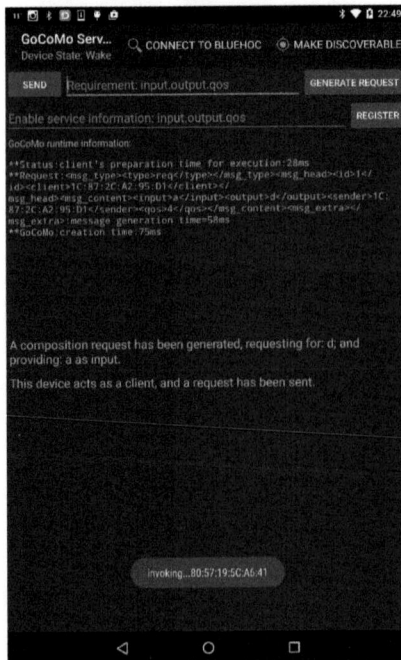

Figure A.3　GoCoMo App: a running device that acts as a composition client

A.2 GoCoMo-ns3

GoCoMo-ns3 works with ns-3 system modules, simulating the GoCoMo composition process in a WiFi ad hoc network. As illustrated in Figure A.4, GoCoMo-ns3 is
used by the **Experiment** class which initializes a number of wireless devices, config-
ures their mobility models, and establishes an ad hoc networks that controls message
interaction between them. GoCoMo-ns3 implemented the **DataReader** class to load a
global service topology, since the ns-3 platform does not manage each individual
node's service information independently.

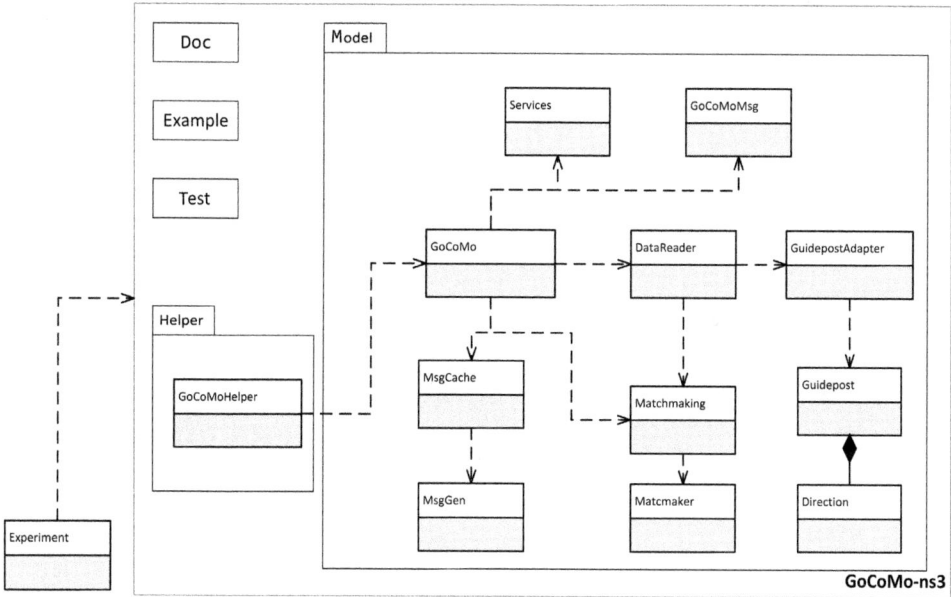

Figure A.4 GoCoMo-ns3's class diagram

Appendix B

Evaluation Results' Validity

This appendix presents the validity of results using 2-sample z-test.

B.1 CoopC and GoCoMo's Service Discovery Time

As 300 (> 30) samples were used for the simulation, this book uses 2-sample z-test. The testable statistical hypothesis for the simulation result is listed below. The null hypothesis (H_0): There is no difference between the population means of Go-CoMo's discovery time and CoopC's discovery time on the test of discovery delay in a service composition process, that is $H_0 : \mu_1 - \mu_2 = 0$.

The alternative hypotheses (H_1 and H_2): There is a difference between the population means of GoCoMo's discovery time and CoopC's discovery time on the test of discovery delay in a service composition process, that is $H_1 : \mu_1 - \mu_2 < 0$ and $H_2 : \mu_1 - \mu_2 > 0$.

The results of z values found through simulations under the null hypothesis and the alternative hypotheses are shown in Table B.1.

Table B.1　The difference between GoCoMo's discovery time and CoopC's discovery time

slow, 5 services	20	30	40	50
z value	−12.56106113	−16.79026592	−18.02333183	−10.69922846
H_0	reject	reject	reject	reject
H_1	accept	accept	accept	accept
H_2	reject	reject	reject	reject
mid-fast, 5 services	20	30	40	50
z value	−19.04224056	−14.76793744	−19.41144055	−12.27153227
H_0	reject	reject	reject	reject
H_1	accept	accept	accept	accept
H_2	reject	reject	reject	reject
fast, 5 services	20	30	40	50
z value	−11.54649281	−14.73933101	−17.47812923	−17.79875104
H_0	reject	reject	reject	reject
H_1	accept	accept	accept	accept
H_2	reject	reject	reject	reject
slow, 10 services	20	30	40	50
z value	−32.29453272	−37.9788997	−26.56567767	−24.23023048
H_0	reject	reject	reject	reject
H_1	accept	accept	accept	accept
H_2	reject	reject	reject	reject

Continued

mid-fast, 10 services	20	30	40	50
z value	−19.30967952	−41.52542078	−31.77308579	−25.29579807
H_0	reject	reject	reject	reject
H_1	accept	accept	accept	accept
H_2	reject	reject	reject	reject
fast, 10 services	20	30	40	50
z value	−19.64695977	−28.91184312	−33.77897592	−24.06613906
H_0	reject	reject	reject	reject
H_1	accept	accept	accept	accept
H_2	reject	reject	reject	reject

B.2   CoopC and GoCoMo's Service Discovery Traffic

The testable statistical hypothesis for the simulation result is listed below.

The null hypothesis (H_0): There is no difference between the population means of GoCoMo's discovery traffic and CoopC's discovery traffic on the test of discovery traffic in a service composition process, that is $H_0 : \mu_1 - \mu_2 = 0$.

The alternative hypotheses (H_1 and H_2): There is a difference between the population means of GoCoMo's discovery traffic and CoopC's discovery traffic on the test of discovery traffic in a service composition process, that is $H_1 : \mu_1 - \mu_2 < 0$ and $H_2 : \mu_1 - \mu_2 > 0$.

The results of z values found through simulations under the null hypothesis and the alternative hypotheses are shown in Table B.2.

Table B.2   The difference between GoCoMo's discovery traffic and CoopC's discovery traffic

slow, 5 services	20	30	40	50
z value	−6.56873	8.410306	−9.09909	−2.9423
H_0	reject	reject	reject	reject
H_1	accept	reject	accept	accept
H_2	reject	accept	reject	reject
mid-fast, 5 services	20	30	40	50
z value	−6.76328	11.75466	−8.38718	−10.0033
H_0	reject	reject	reject	reject
H_1	accept	reject	accept	accept
H_2	reject	accept	reject	reject
fast, 5 services	20	30	40	50
z value	−5.6026	11.03799	−7.44844	−10.00331775
H_0	reject	reject	reject	reject
H_1	accept	reject	accept	accept
H_2	reject	accept	reject	reject

Continued

slow, 10 services	20	30	40	50
z value	−15.4047	−52.666	−53.7117	−39.9217
H_0	reject	reject	reject	reject
H_1	accept	accept	accept	accept
H_2	reject	reject	reject	reject

mid-fast, 10 services	20	30	40	50
z value	−13.5711	−59.179	−45.8739	−46.6946
H_0	reject	reject	reject	reject
H_1	accept	accept	accept	accept
H_2	reject	reject	reject	reject

fast, 10 services	20	30	40	50
z value	−11.2968	−42.1154	−46.6897	−35.1998
H_0	reject	reject	reject	reject
H_1	accept	accept	accept	accept
H_2	reject	reject	reject	reject

B.3　CoopC and GoCoMo's Response Time

The testable statistical hypothesis for the simulation result is listed below.

The null hypothesis (H_0): There is no difference between the population means of GoCoMo's response time and CoopC's response time on the test of response delay in a service composition process, that is $H_0 : \mu_1 - \mu_2 = 0$.

The alternative hypotheses (H_1 and H_2): There is a difference between the population means of GoCoMo's response time and CoopC's response time on the test of response delay in a service composition process, that is $H_1 : \mu_1 - \mu_2 < 0$ and $H_2 : \mu_1 - \mu_2 > 0$.

The results of z values found through simulations under the null hypothesis and the alternative hypotheses are shown in Table B.3.

Table B.3　The difference between GoCoMo's response time and CoopC's response time

slow, 5 services	20	30	40	50
z value	−9.992451711	−8.489276933	−8.994816681	−6.704062264
H_0	reject	reject	reject	reject
H_1	accept	accept	accept	accept
H_2	reject	reject	reject	reject

Continued

mid-fast, 5 services	20	30	40	50
z value	−9.467088577	−8.814763295	−9.297349149	−6.695604027
H_0	reject	reject	reject	reject
H_1	accept	accept	accept	accept
H_2	reject	reject	reject	reject
fast, 5 services	20	30	40	50
z value	−7.663452997	−8.307561975	−7.978716457	−5.495297028
H_0	reject	reject	reject	reject
H_1	accept	accept	accept	accept
H_2	reject	reject	reject	reject
slow, 10 services	20	30	40	50
z value	−22.97000542	−17.12140163	−9.065221592	−3.549828118
H_0	reject	reject	reject	reject
H_1	accept	accept	accept	accept
H_2	reject	reject	reject	reject
mid-fast, 10 services	20	30	40	50
z value	−15.65747282	−16.64296405	−8.09600387	−3.557499945
H_0	reject	reject	reject	reject
H_1	accept	accept	accept	accept
H_2	reject	reject	reject	reject
fast, 10 services	20	30	40	50
z value	−13.60161003	−13.03524901	−9.110352989	−3.6688557
H_0	reject	reject	reject	reject
H_1	accept	accept	accept	accept
H_2	reject	reject	reject	reject

B.4   CoopC and GoCoMo's Composition Traffic

The testable statistical hypothesis for the simulation result is listed below.   The null hypothesis (H_0): There is no difference between the population means of Go-CoMo's composition traffic and CoopC's composition traffic on the test of composition traffic in a service composition process, that is $H_0 : \mu_1 - \mu_2 = 0$.

The alternative hypotheses (H_1 and H_2): There is a difference between the population means of GoCoMo's composition traffic and CoopC's composition traffic on the test of composition traffic in a service composition process, that is $H_1 : \mu_1 - \mu_2 < 0$ and $H_2 : \mu_1 - \mu_2 > 0$.

The results of the simulation found z values under the null hypothesis and the alternative hypotheses are shown in Table B.4.

Table B.4　The difference between GoCoMo's composition traffic and CoopC's composition traffic

slow, 5 services	20	30	40	50
z value	−1.633975083	−0.379535408	−17.01789151	−6.446042073
H_0	accept	accept	reject	reject
H_1	reject	reject	accept	accept
H_2	reject	reject	reject	reject
mid-fast, 5 services	20	30	40	50
z value	−16.25638314	1.10071635	−8.67860778	−25.19971832
H_0	reject	accept	reject	reject
H_1	accept	reject	accept	accept
H_2	reject	reject	reject	reject
fast, 5 services	20	30	40	50
z value	−12.76707972	−0.500782492	−7.814265259	−20.12883122
H_0	reject	accept	reject	reject
H_1	accept	reject	accept	accept
H_2	reject	reject	reject	reject
slow, 10 services	20	30	40	50
z value	−22.75254441	−88.24796791	−56.87641912	−59.74280959
H_0	reject	reject	reject	reject
H_1	accept	accept	accept	accept
H_2	reject	reject	reject	reject
mid-fast, 10 services	20	30	40	50
z value	−10.41214504	−81.90194881	−34.45794557	−55.12380986
H_0	reject	reject	reject	reject
H_1	accept	accept	accept	accept
H_2	reject	reject	reject	reject
fast, 10 services	20	30	40	50
z value	−4.51012531	−64.84092515	−70.28014077	−21.9503516
H_0	reject	reject	reject	reject
H_1	accept	accept	accept	accept
H_2	reject	reject	reject	reject

Appendix C

Glossary of Key Terms

Cache

A cached request sender is represented as a $C_i \in cache$ ($C_i = \langle S_{id}, \mathcal{G}^{matched}, \rho \rangle$), where S_{id} is the unique id of the requester node (e.g., the node Y), and the set $\mathcal{G}^{matched}$ stores matched outputs. The parameter ρ ($\rho \in (0, 1)$) captures the progress of addressing the partially matched goal.

Composition Request

A composition request is represented by $R = \langle R_{id}, I, O, F, C \rangle$, where R_{id} is a unique id for a request. The set $\mathcal{F}$ represents all the functional requirements, which consists of a set of essential while unordered functions. The composition constraints set $\mathcal{C}$ is made up of execution time constraints. A composition process fails if $\mathcal{C}$ expires and the client receives no result during service execution. A service composite request also includes a set of initial parameters (input) $I = \{ \langle I^S, I^D \rangle \}$ and a set of goal parameters (output) $O = \{ \langle O^S, O^D \rangle \}$.

Discovery Message

A discovery message including a request's remaining part R', is represented as $DscvMsg = \langle R, cache, h \rangle$, where $cache$ stores the progress of resolving split-join controls for parallel service flows (See **Definition 6** in section 4.2.1), and h is a criterion value for request forwarding and service allocation (See Section 4.3 and Section 4.4).

Directions

An AND-splitting direction directly links to multiple services, which requires the composite participant to simultaneously invoke these services for execution. An AND-joining direction links to a waypoint-micro service (join-node) that collects data from the composite participant and other services on different branches.

Execution Guidepost

An execution guidepost $G = \langle R_{id}, D \rangle$ maintained by composite participant P includes a set of execution directions $\mathcal{D}$ and the id of its corresponding composite request. For each execution direction $d_j \in \mathcal{D}$, d is defined as $\langle d_j^{id}, S^{post}, \omega, Q \rangle$, where d_j^{id} is a unique id for d_j, and the set S^{post} stores $P's$ post-condition services that can be chosen for next-hop execution. The set ω represents possible waypoints on the direction to indicate execution branches' join nodes when the participant is engaged in parallel data flows. The set Q reflects the execution path's robustness of this direction, e.g., the estimated execution path strength and the execution time (See Section 4.4).

Service

A service is described as $S = \langle S^f, IN, UT, QoS^{time} \rangle$, where S^f represents the semantic description of service S's functionality. $IN = \{ \langle IN^S, IN^D \rangle \}$ and $OUT =$

{ $\langle OUT^S, OUT^D \rangle$ } describe the service's input and output parameters as well as their data types, respectively. For this work, execution time QoS^{time} is the most important QoS criterion as delay in composition and execution can cause failures [23]. A service composition model should select services with a short execution time to reduce delay in execution.

Service Announcement Message

A service announcement message is described as $SA = \langle P_{address}, OUT_p \rangle$, where $P_{address}$ represents the unique address of the service provider, and the OUT_p is the output data that can be provided by the service provider.

Service Provider

A service provider is a service deployment device that has a wrapped functionality exposed through a service interface and could therefore be used remotely as a service.

Bibliography

[1]   Banica L, Stefan C, Hagiu A, et al. Leveraging the microservice architecture for next-generation IoT applications[J]. Scientific Bulletin-Economic Sciences, 2017, 16(2):26-32.

[2]   Weiser M. The computer for the 21st century[J]. Scientific American, 1991, 265(3):94-105.

[3]   Bronsted J, Hansen K M, Ingstrup M. Service composition issues in pervasive computing[J]. IEEE Pervasive Computing, 2010, 9(1):62-70.

[4]   Ibrahim N, Mouel F L. A survey on service composition middleware in pervasive environments[J]. International Journal of Computer Science, 2009,1:1-12.

[5]   Raychoudhury V, Cao J, Kumar M, et al. Middleware for pervasive computing: a survey[J]. Pervasive and Mobile Computing, 2013, 9(2):177-200.

[6]   Boubendir A, Bertin E, Simoni N. NaaS architecture through SDN-enabled NFV: network openness towards web communication service providers[C]. Network Operations and Management Symposium, 2016.

[7]   Kekki S, Featherstone W, Fang Y G, et al. MEC in 5G networks[S/OL]. European Telecommunications Standards Institute (ETSI), (2018-06).

[8]   Zhou L, Rodrigues J J P C, Wang H, et al. 5G multimedia communications: Theory, technology, and application[J]. IEEE Multimedia, 2019, 26(1):8-9.

[9]   Shi W, Dustdar S. The promise of edge computing[J]. Computer, 2016, 49(5):78-81.

[10]   Nguyen T D, Kim Y, Kim D H, et al. A proposal of autonomic edge cloud platform with CCN-based service routing protocol[C]. IEEE 11th International Conference on Cloud Computing (CLOUD), 2018.

[11]   Huang Y D, Zhang J R, Duan J, et al. Resource allocation and consensus of blockchains in pervasive edge computing environments[C]. IEEE 39th International Conference on Distributed Computing Systems (ICDCS), 2019.

[12]   Chiang M, Zhang T. Fog and IoT: an overview of research Opportunities[J]. IEEE Internet of Things Journal, 2016, 3(6):854-864.

[13]   Cheng B, Solmaz G, Cirillo F, et al. Fogflow: easy programming of iot services over cloud and edges for smart cities[J]. IEEE Internet of Things Journal, 2018, 5(2):696-707.

[14]   Khan W Z, Xiang Y, Aalsalem M Y, et al. Mobile phone sensing systems: a sur-

vey[J]. IEEE Communications Surveys & Tutorials, 2013, 15:402-427.

[15] Lane N D, Miluzzo E, Lu Hong, et al. Ad hoc and sensor networks: a survey of mobile phone sensing[J]. IEEE Communications Magazine, 2010, 48(9):140-150.

[16] Perera C, Liu C H, Jayawardena S, et al. Context-aware computing in the internet of things: a survey on internet of things from industrial market perspective[J]. IEEE Access, 2014, 2: 1660-1679.

[17] Mian A N, Baldoni R, Beraldi R. A survey of service discovery protocols in multi-hop mobile ad hoc networks[J]. IEEE Pervasive Computing, 2009, 8(1): 66-74.

[18] Fok C L, Roman G C, Lu C Y. Servilla: a flexible service provisioning middleware for heterogeneous sensor networks[J]. Science of Computer Programming, 2012, 77(6):663-684.

[19] Hibner A, Zielinski K. Semantic-based dynamic service composition and adaptation[C]. IEEE Congress on Services, 2007.

[20] Ibrahim Al-Oqily, Ahmed Karmouch. A decentralized self-organizing service composition for autonomic entities[J]. ACM Transactions on Autonomous and Adaptive Systems, 2011, 6(1):1-18.

[21] Prinz V, Fuchs F, Ruppel P, Gerdes C, et al. Adaptive and fault-tolerant service composition in peer-to-peer systems[C]. IFIP International Conference on Distributed Applications and Interoperable Systems, 2008.

[22] Wang H, Sun H, Yu Q. Reliable service composition via automatic QoS prediction[C]. 2013 IEEE International Conference on Services Computing, 2013.

[23] Groba C, Clarke S. Opportunistic service composition in dynamic ad hoc environments[J]. IEEE Transactions on Services Computing, 2014, 7(4):642-653.

[24] Furno A, Zimeo E. Self-scaling cooperative discovery of service compositions in unstructured P2P networks[J]. Journal of Parallel and Distributed Computing, 2014, 74(10):2994-3025.

[25] Khakhkhar S, Kumar V, Chaudhary S. Dynamic service composition[J]. International Journal of Computer Science and Artificial Intelligence, 2012, 2(3): 32-42.

[26] Liu C Y, Cao J, Wang J. A parallel approach for service composition with complex structures in pervasive environments[C]. IEEE International Conference on Web Services, 2015.

[27] Ren K, Xiao N, Chen J. Building quick service query list using wordnet and multiple heterogeneous ontologies toward more realistic service composition[J].

IEEE Transactions on Services Computing, 2011, 4(3):216-229.

[28] Artail H, Mershad K W, Hamze H. DSDM: A distributed service discovery model for manets[J]. IEEE Transactions on Parallel and Distributed Systems, 2008, 19(9):1224-1236.

[29] Prochart G, Weiss R, Schmid R, et al. Fuzzy-based support for service composition in mobile ad hoc networks[C]. IEEE International Conference on Pervasive Services, 2007.

[30] Pirrò G, Talia D, Trunfio P. A DHT-based semantic overlay network for service discovery[J]. Future Generation Computer Systems, 2012, 28(4):689-707.

[31] Rodriguez-mier P, Mucientes M, Lama M. A dynamic QoS-aware semantic web service composition algorithm[C]. International Conference on Service-Oriented Computing, 2012.

[32] Schuler C, Weber R, Schuldt H, et al. Scalable peer-to-peer process management-the OSIRIS approach[C]. IEEE International Conference on Web Services, 2004.

[33] Mokhtar S B, Preuveneers D, Georgantas N, et al. EASY: efficient semantic Service discovery in pervasive computing environments with QoS and context support[J]. Journal of Systems and Software, 2008, 81(5):785-808.

[34] Zisman A, Spanoudakis G, Dooley J, et al. Proactive and reactive runtime service discovery: a framework and its evaluation[J]. IEEE Transactions on Software Engineering, 2013, 39(7):954-974.

[35] Kalasapur S, Kumar M, Shirazi B A. Dynamic service composition in pervasive computing[J]. IEEE Transactions on Parallel and Distributed Systems, 2007, 18(7):907-918.

[36] Sadiq U, Kumar M, Passarella A, et al. Service composition in opportunistic networks: a load and mobility aware solution[J]. IEEE Transactions on Computers, 2015, 64(8):2308-2322.

[37] Daneshfar N, Pappas N, Polishchuk V, et al. Service allocation in a mobile fog infrastructure under availability and QoS constraints[A/OL]. arXiv.org(2017-06-13).

[38] Cao K, Zhou J, Xu G, et al. Exploring renewable-adaptive computation offloading for hierarchical QoS optimization in fog computing[J]. IEEE Transactions on Computer-Aided Design of Integrated Circuits and Systems, 2019, 39(10): 2095-2108.

[39] Kang E, Kim M, Lee E, et al. DHT-based mobile service discovery protocol for mobile ad hoc networks[C]. The 4th international conference on Intelligent

Computing (ICIC), 2008.

[40] He Q, Yan J, Yang Y, et al. A decentralized service discovery approach on peer-to-peer networks[J]. IEEE Transactions on Services Computing,2013, 6(1):64-75.

[41] Bianchini D, De Antonellis V. On-the-fly collaboration in distributed systems through service semantic overlay[C]. The 10th International Conference on Information Integration and Web-based Applications & Services, 2008.

[42] Bianchini D, De Antonellis V, Michele M. P2P-SDSD: on-the-fly service- based collaboration in distributed systems[J]. International Journal of Metadata Semantics & Ontologies, 2010, 5(3): 222-237.

[43] Orsini G, Bade D, Lamersdorf W. CloudAware: a contextadaptive middleware for mobile edge and cloud computing applications[C]. The 1st International Workshops on Foundations and Applications of Self* Systems (FAS*W), 2016.

[44] Chen L, Pan Z, Gao L, et al. Adaptive fog configuration for the industrial internet of things[J]. IEEE Transactions on Industrial Informatics, 2018, 14(10):4656-4664.

[45] Sen R, Roman G C, Christopher Gill. CiAN: a workflow engine for MANETs[C]. International Conference on Coordination Languages and Models, 2008.

[46] Wang Z, Xu T, Qian Z, et al. A Parameter-based scheme for service composition in pervasive computing environment[C]. International Conference on Complex, Intelligent and Software Intensive Systems, 2009.

[47] Ukey N, Niyogi R, Milani A, et al. A bidirectional heuristic search technique for web service composition[C]. International Conference on Computational Science and Its Applications, 2010.

[48] Gharzouli M, Boufaida M. PM4SWS: A P2P model for semantic web services discovery and composition[J]. Journal of Advances in Information Technology,2011, 2(1):15-26.

[49] Zhu F, Mutka M, Ni L. Splendor: a secure, private, and location-aware service discovery protocol supporting mobile services[C]. The First IEEE International Conference on Pervasive Computing and Communications (PerCom), 2003.

[50] Chakraborty D, Joshi A. GSD: a novel groupbased service discovery protocol for MANETS[C]. The 4th International Workshop on Mobile and Wireless Communications Network, 2002.

[51] Chakraborty D, Joshi A. Toward distributed service discovery in pervasive computing environments[J]. IEEE Transactions on Mobile Computing, 2006, 5(2):97-112.

[52] del Val E, Rebollo M, Botti V. Combination of self-organization mechanisms to enhance service discovery in open systems[J]. Information Sciences, 2014, 279:138-162.

[53] Wang K, Yun S, Xie L, et al. Adaptive and fault-tolerant data processing in healthcare IoT based on fog computing[J]. IEEE Transactions on Network Science and Engineering, 2020, 7(1):263-273.

[54] Kozat U C, Tassiulas L. Service discovery in mobile ad hoc networks: an overall perspective on architectural choices and network layer support issues[J]. Ad Hoc Networks, 2004, 2(1):23-44.

[55] Tyan J, Mahmoud Q H. A comprehensive service discovery solution for mobile ad hoc networks[J]. Mobile Networks and Applications, 2005, 10:423-424.

[56] Kim M J, Kumar M, Shirazi B A. Service discovery using volunteer nodes in heterogeneous pervasive computing environments[J]. Pervasive and Mobile Computing, 2006, 2(3):313-343.

[57] Zhang N, Jian W, He K, et al. An approach of service discovery based on service goal clustering[C]. IEEE International Conference on Services Computing (SCC), 2016.

[58] Zhang N, Jian W, He K, et al. Mining and clustering service goals for RESTful service discovery[J]. Knowledge and Information Systems, 2019, 58(3):1-32.

[59] Tian G, Liu P, Peng Y, et al. Tagging augmented neural topic model for semantic sparse web service discovery[J]. Concurrency and Computation Practice and Experience, 2018, 30(16):e4448.

[60] Crespo A, Garcia-Molina H. Semantic overlay networks for p2p systems[C]. International Workshop on Agents and Peer-to-Peer Computing, 2004.

[61] Furno A, Zimeo E. Efficient cooperative discovery of service compositions in unstructured P2P networks[C]. The 21st Euromicro International Conference on Parallel, Distributed, and Network-Based Processing, 2013.

[62] Thomas L, Wilson J, Roman G C, et al. Achieving coordination through dynamic construction of open workflows[C]. ACM/IFIP/USENIX International Conference on Distributed Systems Platforms and Open Distributed Processing, 2009.

[63] Weißbach M, Chrszon P, Springer T, et al. Decentrally coordinated execution of adaptations in distributed self-adaptive software systems[C]. IEEE 11th International Conference on Self-Adaptive and Self-Organizing Systems (SASO), 2017.

[64] D'Angelo M. Decentralized self-adaptive computing at the edge[C]. The 13th International Conference on Software Engineering for Adaptive and Self-Man-

aging Systems (SEAMS '18), 2018.

[65] Oh S C, Lee D, Kumara S R T. Effective web service composition in diverse and large-scale service networks[J]. IEEE Transactions on Services Computing, 2008 , 1(1):15-32.

[66] Liu X Z, Ma Y, Huang G, et al. Data-driven composition for service-oriented situational web applications[J]. IEEE Transactions on Services Computing, 2015, 8(1):2-16.

[67] Jun T, Roy N, Julien C. Modeling delivery delay for flooding in mobile ad hoc networks[C]. International Conference on Communications (ICC), 2010.

[68] Dai F, Wu J. Performance analysis of broadcast protocols in ad hoc networks based on self-pruning[J]. IEEE Transactions on Parallel and Distributed Systems, 2004, 15(11): 1027-1040.

[69] Geyik S C, Szymanski B K, Zerfos P. Robust dynamic service composition in sensor networks[J]. IEEE Transactions on Services Computing, 2013, 6(4):560-572.

[70] Al Ridhawi Y, Karmouch A. Decentralized plan-free semantic-based service composition in mobile networks[J]. IEEE Transactions on Services Computing, 2015 , 8(1):17-31.

[71] Mokhtar S B, Liu J S. QoS-aware dynamic service composition in ambient intelligence environments[C]. The 20th IEEE/ACM International Conference on Automated Software Engineering, 2005.

[72] De Medeiros R W A, Rosa N S, Pires L F. Predicting service composition costs with complex cost behavior[C]. IEEE International Conference on Services Computing, 2015.

[73] Li Y, Huai J P, Deng T, et al. QoS-aware service composition in service overlay networks. IEEE International Conference on Web Services (ICWS), 2007.

[74] Zhou X, Ge Y, Chen X, et al. A distributed cache based reliable service execution and recovery approach in MANETs[C]. IEEE Asia-Pacific Services Computing Conference, 2011.

[75] Efstathiou D, McBurney P, Zschaler S, et al. Efficient multi-objective optimisation of service compositions in mobile ad hoc networks using lightweight surrogate models[J]. Journal of Universal Computer Science, 2014, 20(8): 1089-1108.

[76] Aschoff R R, Zisman A. QoS-Driven proactive adaptation of service composition[C]. The 9th International Conference on Service-oriented Computing (ICSOC), 2011.

[77] Silas S, Ezra K, Rajsingh E B. A novel fault tolerant service selection framework for pervasive Computing[J]. Human-centric Computing and Information Sciences, 2012, 2(1):1-14.

[78] Ravindra P, Khochare A, Reddy S P, et al. Echo: an adaptive orchestration platform for hybrid dataflows across cloud and edge[C]. International Conference on Service-oriented Computing, 2017.

[79] Groba C, Clarke S. Opportunistic composition of sequentially-connected services in mobile computing environments[C]. IEEE International Conference on Web Services (ICWS), 2011.

[80] Groba C, Clarke S. Synchronising service compositions in dynamic ad hoc environments[C]. IEEE First International Conference on Mobile Services, 2012.

[81] Sousa J P, Poladian V, Garlan D, et al. Task-based adaptation for ubiquitous computing[J]. IEEE Transactions on Systems, Man, and Cybernetics, Part C: Applications and Reviews, 2006, 36(3):328-340.

[82] Vukovi M, Robinson P. Application development powered by rapid , on-demand service composition[C]. IEEE International Conference on Service-Oriented Computing and Applications (SOCA), 2007.

[83] Miraoui M, Tadj C, Fattahi J, et al. Dynamic context-aware and limited resources-aware service adaptation for pervasive computing[J]. Advances in Software Engineering, 2011:1-11.

[84] Fdhila W, Yildiz U, Godart C, et al. A flexible approach for automatic process decentralization using dependency tables[C]. IEEE International Conference on Web Services, 2009.

[85] Chakraborty D, Joshi A, Finin T, et al. Service composition for mobile environments[J]. Mobile Networks and Applications, 2005, 10(4):435-451.

[86] Atluri V, Chun S A, Mukkamala R, et al. A decentralized execution model for interorganizational Workflows[J]. Distributed and Parallel Databases, 2007, 22(1):55-83.

[87] Yu W. Scalable services orchestration with continuation- passing messaging[C]. The First International Conference on Intensive Applications and Services, 2009.

[88] Zaplata S, Hamann K. Flexible execution of distributed business processes based on process instance migration[J]. Journal of Systems Integration, 2010, 1(3): 3-16.

[89] Lipman J, Liu H, Stojmenovic I. Broadcast in ad hoc networks[M]. Misra S, Woungang I, Misra S C. Guide to Wireless Ad Hoc Networks, Computer Com-

munications and Networks. London: Springer, 2009: 121-150.

[90] Karaoglu B, Heinzelman W. Multicasting vs. broadcasting: What are the trade-offs?[C]. IEEE Global Communications Conference (GLOBECOM), 2010.

[91] Sodhro A H, Pirbhulal S, Albuquerque V. Artificial intelligence-driven mechanism for edge computing-based industrial applications[J]. IEEE Transactions on Industrial Informatics, 2019, 15(7):4235-4243.

[92] Lin B, Zhu F, Zhang J, et al. A time-driven data placement strategy for a scientific workflow combining edge computing and cloud computing[J]. IEEE Transactions on Industrial Informatics, 2019, 15(7):4254-4265.

[93] Zhang T, Jin J, Zheng X, et al. Rate-adaptive fog service platform for heterogeneous iot applications[J]. IEEE Internet of Things Journal, 2020, 7(1): 176-188.

[94] Ardagna D, Pernici B. Adaptive service composition in flexible processes[J]. IEEE Transactions on Software Engineering, 2007, 33(6):369-384.

[95] Klusch M. Overview of the S3 contest performance evaluation of semantic service matchmakers[M]. Blake B, Cabral L, König-Ries B, et al. Semantic Web Services. Berlin: Springer, 2012: 17-34.

[96] Cheng D Y, Chao K M, Lo C C, et al. A user centric service-oriented modeling approach[J]. World Wide Web, 2011, 14(4):431-459.

[97] Jiang S S, Xue Y, Schmidt D C. Minimum disruption service composition and recovery in mobile ad hoc networks[J]. Computer Networks, 2009, 53(10):1649-1665.

[98] Jiang S S, Xue Y, Schmidt D C. Minimum disruption service composition and recovery over mobile ad hoc networks[C]. The Fourth Annual International Conference on Mobile and Ubiquitous Systems: Networking & Services (MobiQuitous), 2007.

[99] Gu X H, Nahrstedt K. Distributed multimedia service composition with statistical QoS assurances[J]. IEEE Transactions on Multimedia, 2006 , 8(1):141-151.

[100] Chen N, Yang Y, Zhang T, et al. Fog as a service technology[J]. IEEE Communications Magazine, 2018, 56(11):95-101.

[101] Biswas A R, Giaffreda R. IoT and cloud convergence: opportunities and challenges[C]. IEEE World Forum on Internet of Things (WF-IoT), 2014.

[102] Mell P, Grance T. The NIST definition of cloud computing: special publication (NIST SP) - 800-145 [R/OL]. (2011-09-28).

[103] Sun X, Ansari N. EdgeIoT: mobile edge computing for the internet of things[J]. IEEE Communications Magazine, 2016, 54(12):22-29.

[104] Huber Flores. Edge intelligence enabled by multi-device systems: (percrowd 2020 panel)[C]. IEEE International Conference on Pervasive Computing and Communications Workshops (PerCom Workshops), 2020.

[105] Sasaki K, Suzuki N, Makido S, et al. Vehicle control system coordinated between cloud and mobile edge computing[C]. The 55th Annual Conference of the Society of Instrument and Control Engineers of Japan (SICE), 2016.

[106] Barthélemy J, Verstaevel N, Forehead H, et al. Edgecomputing video analytics for real-time traffic monitoring in a smart city[J]. Sensors, 2019, 19(9):2048.

[107] Okay F Y, Ozdemir S. A fog computing based smart grid model[C]. International Symposium on Networks, Computers and Communications (ISNCC), 2016.

[108] Varghese B, Nan W, Barbhuiya S, et al. Challenges and opportunities in edge computing[C]. IEEE International Conference on Smart Cloud (SmartCloud), 2016.

[109] Bibri S E, Krogstie J. Smart sustainable cities of the future: An extensive interdisciplinary literature review[J]. Sustainable Cities and Society, 2017, 31:183-212.

[110] Yang Y. Multi-tier computing networks for intelligent IoT[J]. Nature Electronics, 2019, 2: 4-5.

[111] Yang Y, Luo X, Chu X, et al. Fog-enabled intelligent IoT systems[M]. Switzerland: Springer, Cham, 2020.

[112] Tran T X, Pandey P, Hajisami A, et al. Collaborative multi-bitrate video caching and processing in mobile-edge computing networks[C]. The 13th Annual Conference on Wireless On-demand Network Systems and Services (WONS), 2017.

[113] Cao N Y, Nasir S B, Sen S, et al. Self-optimizing IoT wireless video sensor node with in-situ data analytics and context-driven energy-aware real-time adaptation[J]. IEEE Transactions on Circuits and Systems I: Regular, 2017, 64(9):2470-2480.

[114] Liang C, Ying H, Yu F R, et al. Enhancing video rate adaptation with mobile edge computing and caching in software-defined mobile networks[J]. IEEE Transactions on Wireless Communications, 2018, 17(10):7013-7026.

[115] Wen Z, Yang R, Garraghan P, et al. Fog orchestration for internet of things services[J]. IEEE Internet Computing, 2017, 21(2):16-24.

[116] Casadei R, Pianini D, Viroli M, et al. Self-organising coordination regions: a pattern for edge computing[C]. International Conference on Coordination Languages and Models, 2019.

[117] Lin W, Lei J, Li J, et al. Online resource allocation for arbitrary user mobility in

distributed edge clouds[C]. IEEE International Conference on Distributed Computing Systems, 2018.

[118] Jie X, Chen L, Ren S. Online learning for offloading and autoscaling in energy harvesting mobile edge computing[J]. IEEE Transactions on Cognitive Communications & Networking, 2017, 3(3):361-373.

[119] Seiger R, Huber S, Heisig P, et al. Toward a framework for self-adaptive workflows in cyber-physical systems[J]. Software & Systems Modeling, 2019, 18: 1117-1134.

[120] Pande V, Marlecha C, Kayte S. A review-fog computing and its role in the internet of things[J]. International Journal of Engineering Research & Applications, 2016, 6(10):7-11.

[121] Agus Kurniawan. Learning AWS IoT: effectively manage connected devices on the AWS cloud using services such as AWS Greengrass, AWS button, predictive analytics and machine learning[M]. Birmingham: Packt Publishing Ltd, 2018.

[122] Chen N, Clarke S. A Dynamic Service Composition Model for Adaptive Systems in Mobile Computing Environments[C]. IEEE International Conference on Service-Oriented Computing (ICSOC), 2014.

[123] Ashrafi T H, Hossain Md A, Arefin S E, et al. Service based fog computing model for IoT[C]. IEEE 3rd International Conference on Collaboration and Internet Computing (CIC), 2017.

[124] Tang B, Chen Z, Hefferman G, et al. Incorporating intelligence in fog computing for big data analysis in smart cities[J]. IEEE Transactions on Industrial informatics, 2017, 13(5):2140-2150.

[125] Chen N, Yang Y, Li J, et al. A fog-based service enablement architecture for cross-domain IoT applications[C]. IEEE Fog World Congress (FWC), 2017.

[126] Garcia S, Cabrera J, Garcia N. Quality-control algorithm for adaptive streaming services over wireless channels[J]. IEEE Journal of Selected Topics in Signal Processing, 2015, 9(1):50-59.

[127] Chen N, Cardozo N, Clarke S. Goal-driven service composition in mobile and pervasive computing[J]. IEEE Transactions on Services Computing, 2018, 11(1):49-62.

[128] Zhao T, Wei Z, Zhao H, et al. A reinforcement learning-based framework for the generation and evolution of adaptation rules[C]. IEEE International Conference on Autonomic Computing (ICAC), 2017.

[129] Tang Z, Zhou X, Zhang F, et al. Migration modeling and learning algorithms for

containers in fog computing[J]. IEEE Transactions on Services Computing, 2018, 12(5):712-725.

[130] Seiger R, Huber S, Heisig P, et al. Enabling self-adaptive workflows for cyber-physical systems[C]. International Conference on Business Process Modeling, Development and Support/International Conference on Evaluation and Modeling Methods of Systems Analysis and Development, 2016.

[131] Ferrández-Pastor FJ, Mora H, Jimeno-Morenilla A, et al. Deployment of IoT edge and fog computing technologies to develop smart building services[J]. Sustainability, 2018, 10(11):1-23.

[132] A. Gatouillat, Y. Badr, and B. Massot. Qos-driven selfadaptation for critical iot-based systems[C]. International Conference on Service-Oriented Computing (ICSOC), 2017.

[133] Kit M, Gerostathopoulos I, Bures T, et al. An architecture framework for experimentations with self-adaptive cyber-physical systems[C]. IEEE/ACM 10th International Symposium on Software Engineering for Adaptive and Self-Managing Systems, 2015.

[134] Weyns D, Ramachandran G S, Singh R K. Self-managing internet of things[C]. International Conference on Current Trends in Theory and Practice of Informatics (SOFSEM), 2018.

[135] Taherizadeh S, Jones A C, Taylor I, et al. Monitoring self-adaptive applications within edge computing frameworks: a state of the art review[J]. The Journal of Systems and Software, 2018, 136:19-38.

[136] Zhang L, Alharbe N, Atkins A S. An IoT application for inventory management with a self-adaptive decision model[C]. IEEE International Conference on Internet of Things (iThings) and IEEE Green Computing and Communications (GreenCom) and IEEE Cyber, Physical and Social Computing (CPSCom) and IEEE Smart Data (SmartData), 2016.

[137] Zolotukhin M, Hamalainen T, Kokkonen T, et al. Increasing web service availability by detecting application-layer DDoS attacks in encrypted traffic[C]. The 23rd International Conference on Telecommunications (ICT), 2016.

[138] Song W, Yin H, Liu C, et al. DeepMem: learning graph neural network models for fast and robust memory forensic analysis[C]. The 2018 ACM SIGSAC Conference on Computer and Communications Security, 2018.

[139] Caporuscio M, D'Angelo M, Grassi V, et al. Reinforcement learning techniques for decentralized self-adaptive service assembly[C]. European Conference on

Service-Oriented and Cloud Computing, 2016.

[140] Mu T Y, Al-Fuqaha A, Shuaib K, et al. SDN flow entry management using rein-forcement learning[J]. ACM Transactions on Autonomous & Adaptive Systems, 2018, 13(2):1-23.

[141] Rahman W U, Hong C S, Huh E N. Edge computing assisted joint quality adap-tation for mobile video streaming[J]. IEEE Access, 2019,7:129082-129084.

[142] Silva R, Fonseca N. Resource allocation mechanism for a fog-cloud infrastruc-ture[C]. IEEE International Conference on Communications (ICC), 2018.

[143] Xiao Y, Jia Y, Liu C, et al. Edge computing security: state of the art and chal-lenges[J]. Proceedings of the IEEE, 2019, 107(8):1608-1631.

[144] Li Q, Zhang Y, Li Y, et al. Capacity-aware edge caching in fog computing net-works[J]. IEEE Transactions on Vehicular Technology, 2020, 69(8):9244-9248.

[145] Xiao Y, Krunz M. Dynamic network slicing for scalable fog computing systems with energy harvesting[J]. IEEE Journal on Selected Areas in Communications, 2018, 36(12):2640-2654.

[146] Chen X, Jiao L, Li W, et al. Efficient multi-user computation offloading for mo-bile-edge cloud computing[J]. I IEEE/ACM Transactions on Networking, 2016, 24(5):2795-2808.

[147] Lin R, Zhou Z, Luo S, et al. Distributed optimization for computation offloading in edge computing[J]. IEEE Transactions on Wireless Communications, 2020, 19(12):8179-8194.

[148] Chen X, Shi Q, Yang L, et al. Thriftyedge: Resourceefficient edge computing for intelligent iot applications[J]. IEEE Network, 2018, 32(1):61-65.

[149]] Liu Q, Wei Y, Leng S, et al. Task scheduling in fog enabled internet of things for smart cities[C]. The 17th International Conference on Communication Tech-nology (ICCT), 2017.

[150] JMinh Q T, Nguyen D T, Le A V, et al. Toward service placement on fog com-puting landscape[C]. The 4th NAFOSTED Conference on Information and Computer Science, 2017.

[151] Jutila M. An adaptive edge router enabling internet of things[J]. IEEE Internet of Things Journal, 2016, 3(6):1061-1069.

[152] Hsieh Y C, Hong H J, Tsai P H, et al. Managed edge computing on inter-net-of-things devices for smart city applications[C]. IEEE/IFIP Network Opera-tions and Management Symposium, 2018.

[153] Gerostathopoulos I, Bures T, Hnetynka P, et al. Self-adaptation in software-intensive

cyberphysical systems: From system goals to architecture configurations[J]. Journal of Systems & Software, 2016, 122: 378-397.

[154] Xiao Y, Shi G, Li Y, et al. Towards self-learning edge intelligence in 6G[J]. IEEE Communications Magazine, 2020, 58(12):34-40.

[155] Zou G, Gan Y, Chen Y, et al. Towards automated choreography of web services using planning in large scale service repositories[J]. Applied Intelligence, 2014, 41(2):383-404.

[156] Nah FFH. A study on tolerable waiting time how long are Web users willing to wait?[C]. Americas Conference on Information Systems (AMCIS), 2003.

[157] Miller R B. Response time in man-computer conversational transactions[C]. The AFIPS Fall Joint Computer Conference, Part I, 1968.

[158] Liu R, Kumar A. An analysis and taxonomy of unstructured Workflows[M]. Aalst W, Benatallah B, Casati F, et al. Business Process Management. Berlin: Springer, 2005.

[159] Sadagopan N, Bai F. PATHS: analysis of PATH duration statistics and their impact on reactive MANET routing protocols[C]. The 4th ACM International Symposium on mobile Ad Hoc Networking & Computing, 2003.

[160] Shaffer J D, Keaveney J F. Automated concierge system and method: US 8,160,614[P]. 2007-02-08.

[161] Breau J R, Miller E E, Ng S Y, et al. Concierge for portable electronic device: US 8,489,080 B1[P]. 2013-07-16.

[162] Mallayya D, Ramachandran B, Viswanathan S. An automaticweb service composition framework using QoS-based web service ranking algorithm[J]. The Scientific World Journal, 2015.

[163] Liu D, Chen B Q, Yang C Y, et al. Caching at the wireless edge: design aspects, challenges, and future directions[J]. IEEE Communications Magazine, 2016, 54(9):22-28.

[164] Bastug E, Bennis M, Debbah M. Living on the edge: the role of proactive caching in 5G wireless networks[J]. IEEE Communications Magazine, 2014, 52(8):82-89.

[165] Shi Y, Chen S, Xu X. Maga: a mobility-aware computation offloading decision for distributed mobile cloud computing[J]. IEEE Internet of Things Journal, 2018, 5(1):164-174.

[166] Zhu Z, Jin S, Yang Y, et al. Time reusing in D2D-enabled cooperative networks[J]. IEEE Transactions on Wireless Communications, 2018, 17(5):3185-3200.

[167] Yang Y, Xu J, Shi G, et al. 5G Wireless Systems[M]. Switzerland: Springer,

Cham, 2018.

[168] Golrezaei N, Molisch A F, Dimakis A G, et al. Femtocaching and device-to-device collaboration: a new architecture for wireless video distribution[J]. IEEE Communications Magazine, 2013, 51(4):142-149.

[169] Chen B Q, Yang C Y, Xiong Z X. Optimal caching and scheduling for cache-enabled D2D communications[J]. IEEE Communications Letters, 2017, 21(5):1155-1158.

[170] Yang Y, Wang K, Zhang G, et al. Meets: maximal energy efficient task schedul-ing in homogeneous fog networks[J]. IEEE Internet of Things Journal, 2018, 5(5):4076-4087.

[171] Yang Y, Wu Y, Chen N, et al. Locass: local optimal caching algorithm with social selfishness for mixed cooperative and selfish devices[J]. IEEE Access, 2018, 6:30060-30072.

[172] Wu Y, Yao S, Yang Y, et al. Challenges of mobile social device caching[J]. IEEE Access, 2016, 4:8938-8947.

[173] Malak D, Al-Shalash M, Andrews J G. Optimizing the spatial content caching distribution for device-to-device communications[C]. IEEE International Sym-posium on Information Theory (ISIT), 2016.

[174] Chen Z Q, Liu Y Y, Bo Zhou, et al. Caching incentive design in wireless D2D networks: A stackelberg game approach[C]. IEEE International Conference on Communications (ICC), 2016.

[175] Zhu K, Zhi W, Zhang L, et al. Social-aware incentivized caching for D2D com-munications[J]. IEEE Access, 2016, 4:7585-7593.

[176] Wang Y, Wu J, Xiao M. Hierarchical cooperative caching in mobile opportunistic social networks[C]. IEEE Global Communications Conference (GLOBECOM), 2014.

[177] Zhu K, Zhi W, Chen X, et al. Socially motivated data caching in ultra-dense small cell networks[J]. IEEE Network, 2017, 31(4):42-48.

[178] Hosny S, Eryilmaz A, Gamal H E. Impact of user mobility on D2D caching net-works[C]. IEEE Global Communications Conference (GLOBECOM), 2016.

[179] Jung K, Park S. Collaborative caching techniques for privacy-preserving loca-tion-based services in peer-to-peer environments[C]. IEEE International Confer-ence on Big Data (Big Data), 2017.

[180] Fawaz K, Abbani N, Artail H. A privacy-preserving cache management system for manets[C]. The 19th International Conference on Telecommunications (ICT),

2012.

[181] Li J, Liu M, Lu J H, et al. On social-aware content caching for D2D-enabled cellular networks with matching theory[J]. IEEE Internet of Things Journal, 2017, 6(1):297-310.

[182] Li Y, Hui P, Jin D P, et al. Evaluating the impact of social selfishness on the epidemic routing in delay tolerant networks[J]. IEEE Communications Letters, 2010, 14(11):1026-1028.

[183] Zhu Y, Xu B, Shi X, et al. A survey of social-based routing in delay tolerant networks: positive and negative social effects[J]. IEEE Communications Surveys & Tutorials, 2013, 15(1):387-401.

[184] Rao J, Feng H, Chen Z Y. Exploiting user mobility for D2D assisted wireless caching networks[C]. The 8th International Conference on Wireless Communications & Signal Processing (WCSP), 2016.

[185] Chen X, Gong X W, Yang L, et al. Exploiting social tie structure for cooperative wireless networking: A social group utility maximization framework[J]. IEEE/ACM Transactions on Networking, 2016, 24(6):3593-3606.

[186] Rosen J B. Existence and uniqueness of equilibrium points for concave n-person games[J]. Econometrica: Journal of the Econometric Society, 1964, 33(3):520-534.

[187] Goebel K, Kirk W A. Topics in metric fixed point theory[M]. Cambridge: Cambridge University Press, 1990.

[188] Burnette Ed. Hello, Android: introducing Google's mobile development platform, 2nd edition[M]. Raleigh: Pragmatic Bookshelf, 2009.

[189] Butler M. Android: changing the mobile landscape[J]. IEEE Pervasive Computing, 2011, 10(1):4-7.

[190] Okediran O O. Mobile operating systems and application development platforms : a survey[J]. International Journal of Advanced Networking & Applications, 2014, 6:2195-2201.

[191] Gronli T M, Hansen J, Ghinea G, et al. Mobile application platform heterogeneity: Android vs windows phone vs iOS vs firefox OS[C]. IEEE 28th International Conference on Advanced Information Networking and Applications, 2014.

[192] Pandey M, Nakra N. Consumer preference towards smartphone brands, with special reference to Android operating system[J]. IUP Journal of Marketing Management, 2014, 13(4):7-22.

[193] AppBrain. Number of Android applications[A/OL]. AppBrain. (2015).

[194] Pigadas V, Doukas C, Plagianakos V P, et al. Enabling constant monitoring of chronic patient using Android smart phones[C]. The 4th International Conference on Pervasive Technologies Related to Assistive Environments (PETRA'11), 2011.

[195] Postolache O, Girao P S, Ribeiro M, et al. Enabling telecare assessment with pervasive sensing and Android OS smartphone[C]. IEEE International Symposium on Medical Measurements and Applications, 2011.

[196] Hinojos G, Tade C, Park S, et al. BlueHoc : Bluetooth ad hoc network Android distributed computing[C]. The International Conference on Parallel and Distributed Processing Techniques and Applications, 2014.

[197] Ns-3. ns-3 Tutorial: A discrete-event network simulator[A/OL]. 2015.

[198] Flyvbjerg B. Five misunderstandings about case-study research[J]. Qualitative inquiry, 2006, 12(2):219-245.

[199] Wohlin C, Höst M, Henningsson K. Empirical research methods in software engineering[M]// Empirical Methods and Studies in Software Engineering. Berlin, Heidelberg: Springer, 2003, 2765:7-23.

[200] Wohlin C, Runeson P, Höst M, et al. Experimentation in software engineering[M]. Boston: Springer, 2000.

[201] Kitchenham B, Pickard L, Pfleeger S L. Case studies for method and tool evaluation[J]. IEEE Software, 1995, 12(4):52-62.

[202] Kiess W, Mauve M. A survey on real-world implementations of mobile ad hoc networks[J]. Ad Hoc Networks, 2007, 5(3):324-339.

[203] Mateescu R, Poizat P, Salaün G. Adaptation of service protocols using process algebra and on-the-fly reduction techniques[C]. International Conference on Service-Oriented Computing, 2008.

[204] Bucchiarone A, Kazhamiakin R, Cappiello C, et al. A context-driven adaptation process for service-based applications[C]. The 2nd International Workshop on Principles of Engineering Service- Oriented Systems (PESOS '10), 2010.

[205] Mostarda L, Marinovic S, Dulay N. Distributed orchestration of pervasive services[C]. The 24th IEEE International Conference on Advanced Information Networking and Applications, 2010.

[206] Poizat P, Yan Y H. Adaptive composition of conversational services through graph planning encoding[C]. International Symposium On Leveraging Applications of Formal Methods, Verification and Validation, 2010.

[207] Newman P, Kotonya G. A Runtime resource-aware architecture for service-oriented

embedded systems[C]. Joint Working IEEE/IFIP Conference on Software Architecture and European Conference on Software Architecture, 2012.

[208] Hogie L, Bouvry P, Guinand F. An overview of MANETs simulation[J]. Electronic Notes in Theoretical Computer Science, 2006, 150(1):81-101.

[209] Liu Y, Han Y Y, Yang Z P, et al. Efficient data query in intermittently- connected mobile ad hoc social networks[J]. IEEE Transactions on Parallel and Distributed Systems, 2015, 26(5):1301-1312.

[210] Efstathiou D, McBurney P, Zschaler S, et al. Efficiently approximating the QoS of composite services in mobile ad-hoc networks[R]. Technical Report, 2014.

[211] Breslau L, Estrin D, Fall K, et al. Advances in network simulation[J]. Computer, 2000, 33(5):59-67.

[212] Hiyama M, Kulla E, Ikeda M, et al. A Comparative study of a MANET testbed performance in indoor and outdoor stairs environment[C]. The 15th International Conference on Network-Based Information Systems, 2012.

[213] Kumar C N, Sam R P. MANET test bed for rescue operations in disaster management[J]. International Journal of Computer Science and Mobile Computing, 2015, (7):432-437.

[214] Niida S, Uemura S, Nakamura H. User tolerance for waiting time[J]. Vehicular Technology Magazine, 2010,5(3):61-67, 2010.

[215] Kotz D, Newport C, Gray R S, et al. Experimental evaluation of wireless simulation Assumptions[C]. The 7th ACM International Symposium on Modeling, Analysis and Simulation of Wireless and Mobile Systems (MSWiM), 2004.

[216] Li B, Pei Y J, Wu H, et al. Heuristics to allocate high-performance cloudlets for computation offloading in mobile ad hoc clouds[J]. The Journal of Supercomputing, 2015, 71(8):3009-3036.

[217]Lee J S, Su Y W, Shen C C. A comparative study of wireless protocols: bluetooth, UWB, ZigBee, and Wi-Fi[C]. IEEE Conference Industrial Electronics Society, 2007.

[218] Kiepuszewski B, Harry A, Bussler C J. On structured workflow modelling[C]. International Conference on Advanced Information Systems Engineering (CAiSE), 2000.

[219] Newman M E J, Watts D J, Strogatz S H. Random graph models of social networks[J]. Proceedings of the National Academy of Sciences, 2002, 99(1): 2566-2572.

[220] Wang X F, Chen G. Complex networks: smallworld, scale-free and beyond[J]. IEEE Circuits and Systems Magazine, 2003, 3(1):6-20.

[221] Scott J, Gass R, Crowcroft J, et al. Crawdad the Cambridge/haggle dataset[DS/OL]. crawdad (2009-05- 29)(2009).